AF650044

Pesticide application equipment for use in agriculture

Vol. 1
Manually carried equipment

by
G.A. Mathews
E.W. Thornhill

United Book Prints

Reprinted in India by:

UNITED BOOK PRINTS
(Imprint of Scientific Publishers, INDIA)
4806/24, Ansari Road, Daryaganj,
New Delhi 110 002, INDIA
E-mail: info@scientificpub.com
Website : http://www.scientificpub.com

Print : 2026

The designations employed and the presentation of material in this publication do not imply the expression of any opinion whatsoever on the part of the Food and Agriculture Organization of the United Nations concerning the legal status of any country, territory, city or area or of its authorities, or concerning the delimitation of its frontiers or boundaries.

All rights reserved. No part of this publication may be reproduced, stored in a retrieval system, or transmitted in any form or by any means, electronic, mechanical, photocopying or otherwise, without the prior permission of the copyright owner. Applications for such permission, with a statement of the purpose and extent of the reproduction, should be addressed to the Director, Publications Division, Food and Agriculture Organization of the United Nations, Viale delle Terme di Caracalla. 00100 Rome. Italy.

ISBN: 978-93-83692-76-7

© **FAO** 1994

Printed in India

FOREWORD

Pesticides are an important input worldwide for crop production. The effectiveness of pesticide use, however, is greatly influenced by inadequate or poor application techniques, which can also lead to serious environmental and health problems, both animal and human.

Studies carried out by PAO have shown a big discrepancy between the available "state of the art" technology which is now commercially available, and the plant protection equipment and application practices often found at field level. This harmful and sometimes tragic discrepancy originates mostly from a vel:y limited knowledge on the subject of pesticide application within the farming, governmental and commercial sectors and is especially serious in formerly centrally planned economies of eastern and central europe and developing countries.

As a result of these studies, PAO clecicleclto update its technical literature. This Agricultural Services Bulletin on manually•carrieclpesticide application equipment is the first of two new volumes replacing Bulletin 38, first published in 1979, which has become very outdated.

The Bulletin has been compiled to improve the technical knowledge on pesticide application equipment and its use and incorporates the latest advances in technology. It is for use by agricultural engineers, technicians, and training establishments as well as casual readers wishing to broaden their knowledge of the subject.

The present first Volume will b_e complemented by a second one on mechanically powered pesticide application equipment for use in agriculture.

The bulletin was prepared under contract for PAO by G.A. Matthews and E.W. Thornhill at the International Pesticide Application Research Centre of Imperial College, Silwood Park, Great Britain. Contributions and comments have been provided by many colleagues both from inside and outside FAO. Among those to whom special thanks are owed are: K. Eveleens (University of Wageningen/The Netherlands), D. Giga (University of Zimbabwe), P.-W. Kopisch-Obuch (PAO/Rome), W. Uicke (University of Gcittingen/Germany) and L. Lumkes (The Netherlands).

AclrianusG. Rijk, Chief
PAO Agricultural Engineering Service.

Contents

Abstract

This Bulletin describes the main types of manually-operated or manually-carried pesticide application equipment.· Most of the small-scale farmers in the developing countries, who have used pesticides, rely on a lever operated knapsack sprayer. These sprayers are now manufactured in many of the developing countries, who are introducing modern designs. The details of the design and the types of nozzle needed to treat crops and improve safety to the user are described. Changes in manufacturing techniques are resulting in certain plastics being used instead of brass.

The less expensive compression sprayers are also described as these may become more important on the smallest farms. Details of the range of nozzles that can be used with hydraulic sprayers are given so the user can make a better selection of nozzle to apply pesticides more efficiently. Equipment used to apply reduced volumes such as the spinning disc applicators is important, especially in areas where water supplies are difficult to obtain. Fog, granule, dust and other special applicators are also described.

Specifications for the main types of applicator are given to assist those who have to purchase the equipment. The remaining sections provide guidance on safe application of pesticides and maintenance oft he equipment.

A glossary of terms is included.

Key words

Pesticide, Application Equipment, Specifications, Knapsack Sprayers, Compression Sprayers, Spinning Discs Applicators, Motorised Mistblowers, Safety.

Whenever using a pesticide

ALWAYS READ THE LABEL

of the Product to ensure that

all the recommendations

of the Manufacturer are carried out.

CHAPTER 1

THE ROLE OF PESTICIDES IN INTEGRATED PEST MANAGEMENT AND BASIC FACTORS AFFECTING CHEMICAL CONTROL

The world's human population is continuing to increase, especially in many tropical areas. By the year 2000, there will be an estimated population of over 6,000 million. To feed and clothe these people, food crop yields need to be increased and more natural fibres produced, preferably near to the main centres of population increase to minimise the need for transport and foreign exchange, and to keep people employed in agriculture rather than increase the problems of urbanisation.

Higher yields of most of the major crops have been obtained by selecting new cultivars, some with significant levels of resistance to some pests. However, achieving the potential yield of these cultivars over large areas has also depended to a large extent on increasing inputs, such as fertilizer, and ensuring effective weed, insect and disease control. Over the last 40 years, the farmer has been able to select and purchase from a wide range of pesticides to control the many different pests that can attack crops. However, there is increasing concern that many of these chemicals have adverse effects on the environment. When it was used widely, DDT and some of the other organochlorine compounds being highly persistent reduced survival of many types of birds due to accumulation in food chains. When less persistent organophosphate insecticides were used instead of DDT, there were more problems of human health as these were more toxic to man.

To-day the emphasis is on integrated pest management to minimise the use of pesticides by also using other control tactics, cultural and biological, in as harmonious a way as possible to keep the numbers of pests so low that an application of a pesticide is not needed. Some natural enemies can be purchased and released, sometimes on several occasions, like an insecticide, on certain crops. In these situations, pesticide use must be limited and selective to avoid interfering with the biological control.

The agrochemical industry's main market has been the larger farms in the United States of America and western Europe. Similarly in the developing countries the main use of pesticides has been on large commercial estates, for example the application of herbicides in oil palm, rubber and coffee plantations, fungicides on coffee and insecticides mainly on cotton. In each of these estates relatively few people purchase large quantities of pesticide. In contrast, marketing to the vast number of poor resource tropical farmers is very difficult and expensive as each of them, who wishes to use a pesticide,only requires small packs. This has meant that where pesticides have been used on these small farms, it has been mainly due to a large-scale Government or aid-sponsored scheme.

However, the market for pesticides in the USA and Europe is no longer growing and may decline as there is over-production of some agricultural produce. In consequence the agrochemical industry is now recognising that any growth in the pesticide industry will now be more likely in the developing coun.tries.

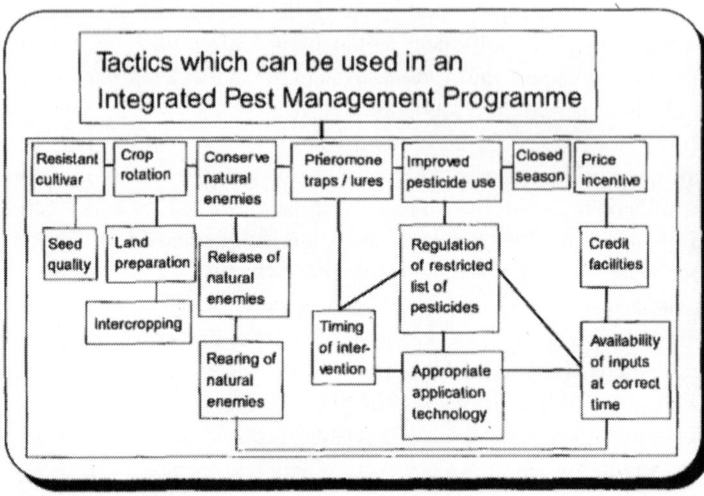

The small size of many farms in the tropics, the type of terrain for crops, like coffee, grown on hillsides, and the lack of capital to invest in large equipment, are some of the factors which result in most of the small-scale growers, who adopt chemical control, requiring manually-operated, or at least manually-carried equipment. At present most of the small-scale farmers using pesticides have a lever-operated knapsack sprayer. Over 60 million sprayers are probably in use world-wide, but few of the farmers are educated technically and in many cases the quality of the application is poor. Quality of equipment varies and poor upkeep results in leaks contaminating the user and worn nozzles giving poor spray coverage. This Bulletin concentrates on a description of the types of equipment that might be used by individual farmers with a small farm. The information is intended to assist advisers and trainers to pass on information to farmers. It indicates the type of Specification required so that Tender Boards and other purchasers of equipment have an outline of the technical requirements needed, rather than just purchase on the basis of price alone. Safety during application of pesticides is considered in detail and advice given on how the main types of equipment can be maintained.

Application of different types of pesticide

The very wide range of crops grown by small-scale farmers and the warm or hot climates favourable for pests means that most tropical farms suffer some losses due to pests, diseases and weeds. Therefore the farmer adopting chemical control may need to apply one or more different types of pesticides. Some act by CONT ACT with the pest, while other are absorbed into the crop plants and are referred to as SYSTEMIC (moving up the plant from the roots or lower leaves and stem) or TRANSLOCATED (some move down the plants such as glyphosate into the rhizome) pesticides. As some pests are on the undersurface of leaves, a insecticide with a TRANSLAMINAR action or VAPOUR action is needed. Some insecticides have to be INGESTED by the insect, ie STOMACH poisons. Many of the more recent pesticides are active with only a few grams compared with kilograms per hectare for older products, and thus greater accuracy in application is very important.

Insecticides

Many different types of chemical insecticides are now available, including some products of biological origin such as natural pyrethrins. In relation to their use in Integrated Pest Management programmes, the more selective insecticides, and least hazardous should be selected. Some of the newest insecticides affect the growth and development of insects and are less hazardous to parasitoids and predators. These include the acyl-urea insecticides or insect growth regulators (IGRs). Examples of these include diflubenzuron, teflubenzuron, flufenoxuron, lufenuron and buprofezin. Most of these are effective against the larval stages of moths, such as cutworms and armyworms, but buprofezin has been used mainly in rice against the brown planthopper.

The delta-toxin derived from *Bacil/l1sllmril1giel1sis* has now been extensively used as a stomach poison against various pests, for example early season cotton leaf-feeders, such as *Alabama argillctcea,* without detrimental effects on natural enemies. Different strains of Bt. have been made available commercially. As this insecticide has to be ingested it is essential to deposit the spray where the insects feed. Droplet size can be impo1iant to get retention of the surface yet carry sufficient Bt that a toxic dose is transferred without inhibiting feeding.

Other selective sprays used on some crops include soaps for aphid control and mineral oil emulsions to restrict virus transmission by whiteflies and other sucking pests.

Broad-spectrum chemicals, such as most organophosphate and pyrethroid insecticides will continue to be needed but they also kill beneficial insects. Thus, when these have to be applied, the application should limit the distribution of the pesticide as much as possible to the part of the crop most exposed to the pest. With insecticides always avoid spray contaminating untreated crops and hedgerows downwind, so that non-target species can survive. This is very important in inter-cropping areas, as the untreated crops may be an important refuge for beneficial species. Always try to avoid treating a very extensive area simultaneously so that there areas where beneficial species can survive and re-invade the treated area.

Fungicides

The main use of fungicides in the tropics has been with crops such as coffee to control leaf rust, but increasingly fungicides have been used on many vegetable and flower crops. Application of the older contact fungicides such as copper sprays requires careful directed application to the sites requiring protection. New growth or loss of spray deposits due to rain may require repeat applications at short intervals. Spray distribution of systemic fungicides may be less critical but good coverage can minimise the dosage required.

Herbicides

The application of herbicides on small-scale farms is much more difficult for the farmer, who may grow several different crops in the same field, as downwind drift of spray droplets must be minimised. Similarly the farmer needs to avoid using a herbicide which persists too long, if a different crop is sown next in the same land. Some SELECTIVE herbicides can be applied direct to the weeds and crop, while careful placement of other BROAD-SPECTRUM herbicides is needed to avoid crop damage. Herbicides may be applied PRE-EMERGENCE before the crop or weeds have appeared or applied as a POST-EMERGENT spray direct to the weeds. It is important with post-emergence treatments that weeds are controlled at the appropriate growth stage. In most cases the weed should be actively growing and as small as possible to reduce the amount of herbicide required.

Nematicides

Nematodes, such as root knot nematode, are small worms, which attack the roots of crops. Tobacco and tomatoes are two crops which can be heavily infested by nematodes. Crops need to be rotated with a long period of 5 years between susceptible crops being grown on the same piece of land. This is often not possible so if the small farmer decides to use chemical control, there are some granular nematicides. Unfortunately these contain some of the most toxic pesticides, such as aldicarb, and require very careful and accurate application. Incorporation into the soil is important to the depth at which the roots of a crop will be growing.

Biologicals

Pathogens

Concern about the pollution of the environment with persistent broad-spectrum pesticides has led to increased research on the use of biological control agents. Some of these, for example certain baculo-viruses and fungal pathogens, have been formulated so that they can be applied with the same equipment used for chemical pesticides. *Beauveria hassiana* is one of the mycopesticides that has been used in several countries. The number of successes has been limited, and in some cases this is due to the lack of the environmental conditions for the biological agent to become established. Correct timing of an application is often crucial.

When *Verticillium lecanii* is used against aphids on chrysanthemums a very high humidity is required and this is obtained by enclosing the plants in a polythene cover, needed also to control day length and initiate flowering.

Experiments with *lvfetarrhizium .flavoviride* to control locusts and grasshoppers have used an oil-based formulation applied at ultra-low volume so that spores are delivered to the insects without drying out. Droplet size is considered to be important to improve collection of the spores on insects, so a spinning disc applicator (see Chapter 6) is recommended.

Therefore it is likely that in the development of other mycopesticides formulation and application techniques will have to be adapted to the specific requirements of the pathogen, pest and crop.

Parasitoids and predators

Some natural enemies are available commercially for release in crops. Predatory mites and the parasitoid *Encarsia formosa* have been widely used on horticultural crops to control red spider mites and whiteflies. Some of the natural enemies are attached to cards that can be hung in a crop, while others are mixed with vermiculite or alternative inert material so that packages of natural enemies can be sent to growers by post. The particles with the natural enemies can be dispersed by sprinkling them through the crop, but it is expected that equipment to disperse natural enemies will also be developed to reduce the time required for manual releases.

Formulation of Pesticides

Pesticide manufacturers have developed a range of formulations to facilitate application. Some of these are sold as liquids or suspensions which can be poured into a measuring container for transfer to the sprayer. However, as this is the stage at which the user is at most risk of being contaminated by the concentrated product, there is increasing use of new packaging systems, such as incorporating a measure in the container or by using direct transfer systems. Other formulations are sold as a water-dispersible granule or wettable powder, which are mixed with water before use. These dry products are difficult to measure in small quantities so some are sold in pre-measured water soluble sachets or pelleted into a tablet that can be added directly to the sprayer.

Great care is always needed in measuring the correct quantity of pesticide for each application. Overuse is a problem where farmers add a little more to make sure it will be effective. Too much can be phytotoxic and damage leaves.

Some pesticides are applied as dry formulations - granules and dusts. These are considered in Chapter 7.

Figure I Container with in-built measure.

Formulation Codes

A two-letter code system has been adopted internationally to indicate the types of formulation. The main formulations are as follows:

A) Products supplied as concentrates for dilution with water.

EC	emulsifiable concentrate
WP	wettable powder
SC	suspension concentrate
WG	water dispersible granule
SG	water soluble granule

B) Products to be applied without any dilution

RB	bait
GR	granules
DP	dusts
UL	ultra-low volume spray liquids
ED	electrochargeable liquid
HN	concentrate suitable for use with thermal fogger
KN	concentrate for cold fogger
FU	smoke generator

Biological Target

The trend in the past has been to treat the whole area of the crop, namely the field. However, pests are often confined to one part of the field or one section of plants, the size of which will change during the season. To reduce the amount of pesticide applied as much as possible, the spray should be confined to patches within a field or to specific sites with a crop canopy. With small manually operated equipment, this is much easier to do than with large mechanised sprayers.

Small droplets (<100μm diameter) are more likely to move between layers of leaves and be filtered out by insects than the largest droplets (>300μm) that fall mainly due to gravity. These larger droplets are either retained on the outer upper leaves of crops, or fall through the canopy to lower leaves or the soil under the crop. The differences in the movement of droplets due to their size can be exploited to improve coverage (droplets per unit area) with small droplets or reduce downwind drift by using large droplets. Directing nozzles at an angle to the crop rather than vertically downwards will increase deposits on the upper canopy and can reduce soil contamination. With hydraulic nozzle sprayers which produce a range of droplet sizes, pressure control is considered to be very important and is discussed later. On some other types of sprayer the size of droplets can be selected and this is considered in Chapter 6.

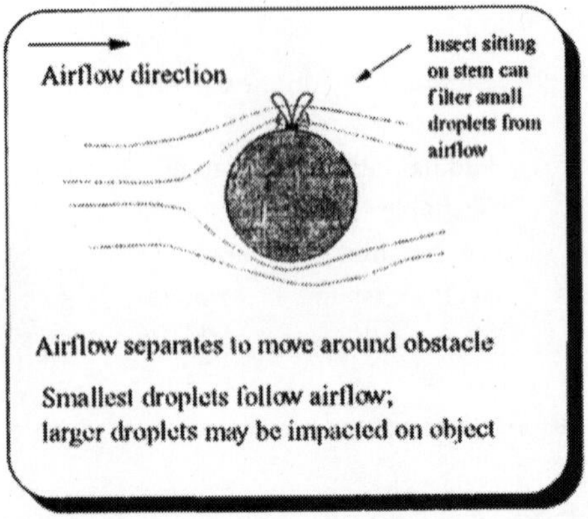

Another factor that should be carefully considered is the dose of pesticide applied. The dose recommended on the label is intended to control pests under most circumstances, including when the farmer applies an application late. The dose of insecticide to control large larvae chewing a crop will be much greater than if the young stages are controlled. Thus crop monitoring is important to detect the early stages of a pest, especially if the initial infestation is restricted to small areas within a crop. Large dosages applied to a crop increase the persistence of deposits, and could select for resistance to the pesticide quicker than when low, less persistent deposits are on the

Spray Volumes

Water has to be collected and carried often over long distances from a borehole or stream to the field that needs treatment. In order to minimise the transportation of large quantities of water, which can take many hours, the farmer needs to keep the volume of spray applied as low as possible. Unfortunately the small nozzles used on hydraulic nozzle/pressure sprayers are most liable to blockage so a compromise is needed. For most hydraulic nozzle sprayer applications on small-scale farms, the volume of spray should be between 100 and 250 litres per hectare. Lower volumes require special equipment and are discussed in Chapter 6. Higher volumes are generally impractical in terms of the time and effort needed just for obtaining a supply of water. They are also wasted as the large droplets that impinge on the crop will coalesce and drip off the leaves. Another factor is that if a pesticide is diluted too much the surfactant in the spray will be too weak to get adequate wetting of certain foliage with a waxy surface.

Table 1 Time taken to spray one hectare with different volume rates.

Volume per hectare	*400*	*200*	10	*1*
sprayer tank litres	15	15		1
mixing/ refill time (h/ha)	2.2	1.1	0.67	0.07
Ferry time to collect water (h/ha)	14.8	7.4	0.37	0.04
Spraying time (h/ha)	2.8	2.8	2.8	2.8
Total time (h/ha)	19.9	11.4	4.0	1
Actual spray time as percentage of total time required	14	25	71	60

Whenever possible label recommendations should indicate the recommended volume of application in litres per hectare, rather than give general terms such as low, medium or high volume. These terms have different interpretations according to the organisation making recommendations and in relation to field and tree crops.

Timing of Application

Frequently a pesticide application has been recommended on a calendar schedule with fixed dates, irrespective of the pest infestation. While some pesticides such a pre-emergence herbicides require a specific timing, the aim should always be to assess pest numbers and then decide if a pesticide treatment is needed. As noted previously, treatment of the younger nymph or larval stages may require less pesticide than when there- are adults or large larvae. The first instar larvae between egg hatch and penetration into the plant is often the only stage when larvae are exposed to leaf surface deposits of contact insecticides. The difficulty is that farmers may not see a pest infestation until it is well established and then it is more difficult to control. Routine crop inspections should be made to recognise pest outbreaks at an early stage. This may also allow a smaller dosage to be effective or for treatment to be localised. Spraying in hot sunlight may be a cause of phytotoxicity.

Environmental Factors

The effect of the wind, humidity and weather conditions generally on the effectiveness of a pesticide application are now more fully understood. In particular a strong wind can carry spray droplets or dry particles over long distances if their mass is insufficient for gravity to pull them down onto the crop, weed or soil. The speed at which different sized water droplets would fall in still air is as follows:

Droplet size μm	*Fall velocity* 111/s
200	0.721
100	0.279
50	0.075
10	0.003

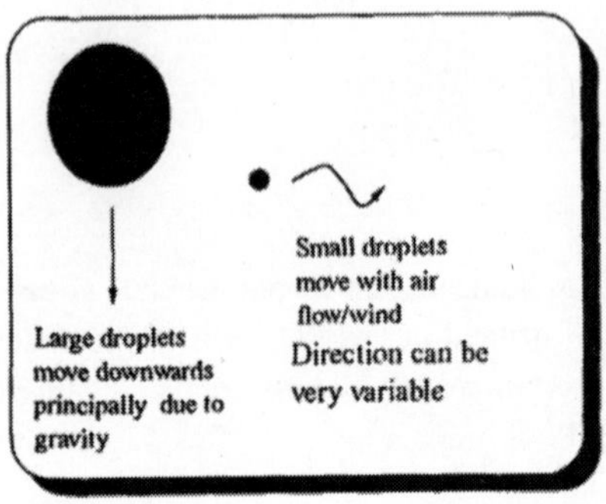

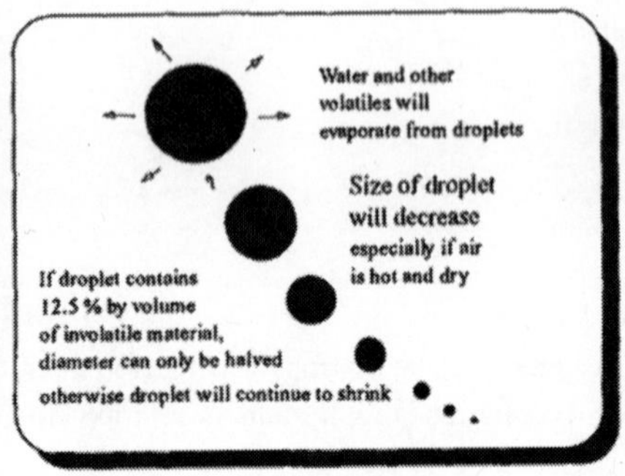

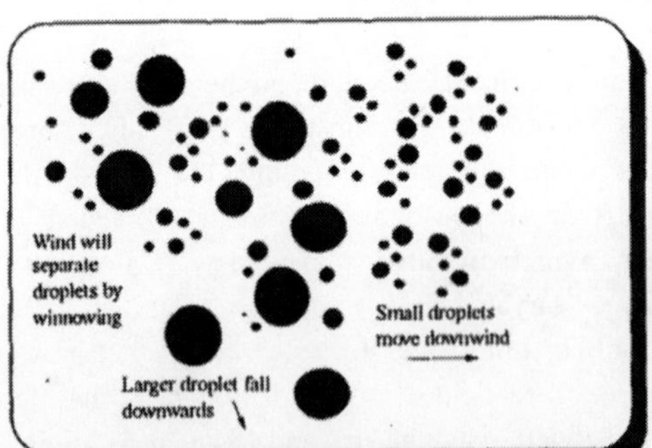

However, water evaporates from these droplets, while they are in flight. The rate of evaporation increases as the droplets get smaller with a greater surface area, and when temperatures ~re high and in dry air when the humidity is low. A droplet of only water will become water vapour very quickly. Thus the lifetime of droplets in a tropical climate such as 30°C and 50% relative humidity is approximately as follows:

Droplet size *µlll*	*Ltfetime of droplet* *seconds*
200	56
100	14
50	3.5
10	0.16

A droplet of pesticide spray contains the active ingredient and other components of the formulation, so if all the water and other volatile liquid has gone, the remaining pesticide particle is now aerosol sized.

The distance downwind that a droplet will travel will increase as the droplet size and its fall velocity decreases. The largest droplets with a high fall velocity will travel almost vertically on to the nearest surface. In contrast, the smallest droplets, particularly aerosols, have the potential to travel extremely long distances. However, the airflow over a field is affected by the smoothness or otherwise of its surface and the effects of soil and air temperature.

Early in the morning, there is an 'inversion', when temperatures in or above a crop are higher than those at the soil surface. Under these conditions the air is usually very calm, and if small droplets are sprayed at this time they can remain airborne for a long time. The ultimate destination of the smallest droplets is unpredictable, when there is no wind to blow them in one direction. With. no wind, droplets eventually settle gravity downwards and SEDIMENT on the upper surface of horizontal surfaces. There have been occasions in Africa where a herbicide applied to a roadside on a hill adversely affected a crop, which was sensitive to that particular chemical, in a valley several kilometres away from where the herbicide was applied.

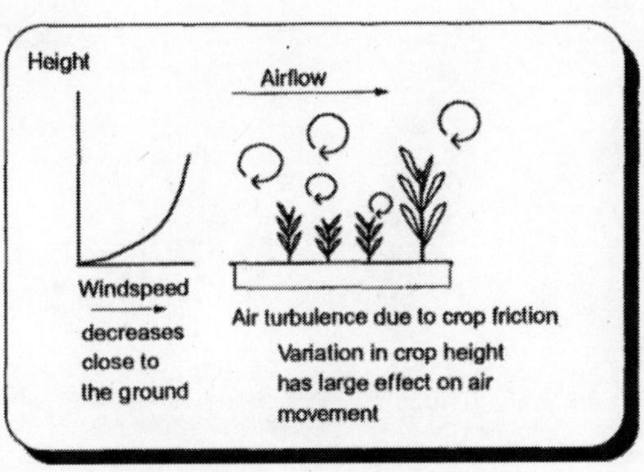

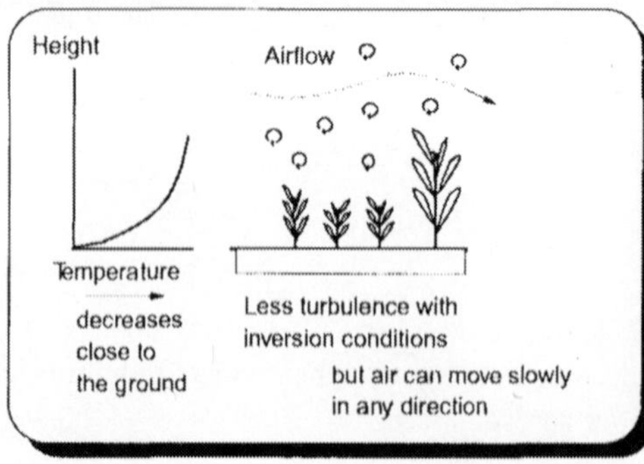

Wind speeds

Approximate airspeed at one metre	Beaufort scale at 10 metres	Description	Visible signs	Spraying
less than 2 km/h	Force 0	Calm	Smoke rises vertically	Avoid spraying
2 - 3.2 km/h	Force 1	Light air	Smoke drifts in one direction	Avoid spraying
3.2 - 6.5 km/h	Force 2	Light breeze	Leaves rustle	Ideal spraying
6.5 - 9.6 km/h	Force 3	Gentle breeze	Leaves in constant motion	Avoid herbicide sprays
9.6 - 14.5 km/h	Force 4	Moderate	Raises dust	Spraying not advised

A simple wind meter for indicating whether conditions are suitable for spraying.

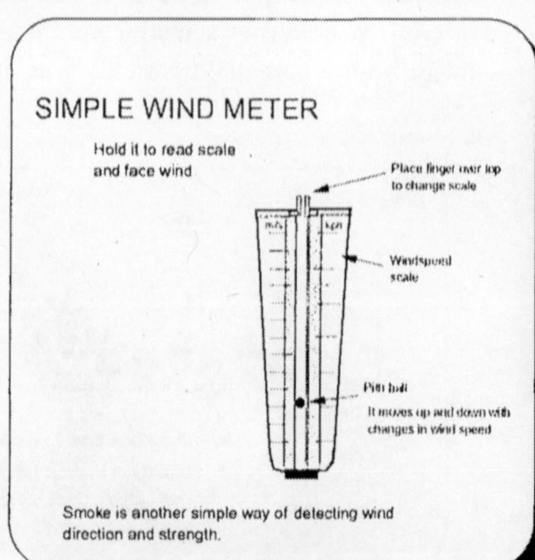

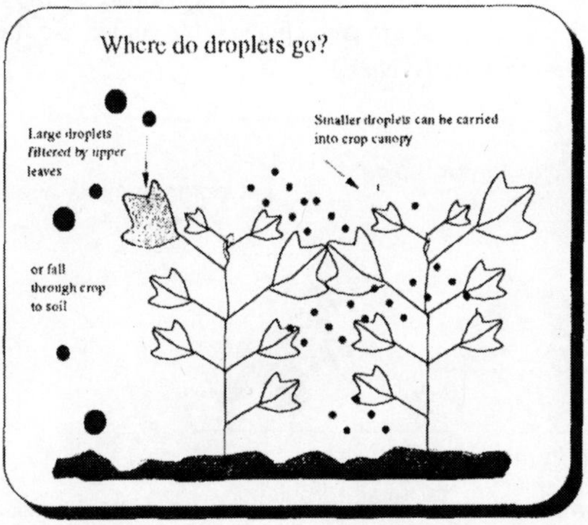
Where do droplets go?
Large droplets filtered by upper leaves
Smaller droplets can be carried into crop canopy
or fall through crop to soil

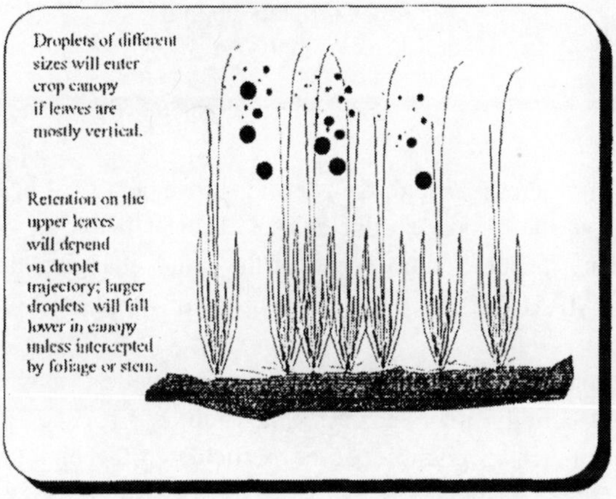
Droplets of different sizes will enter crop canopy if leaves are mostly vertical.
Retention on the upper leaves will depend on droplet trajectory; larger droplets will fall lower in canopy unless intercepted by foliage or stem.

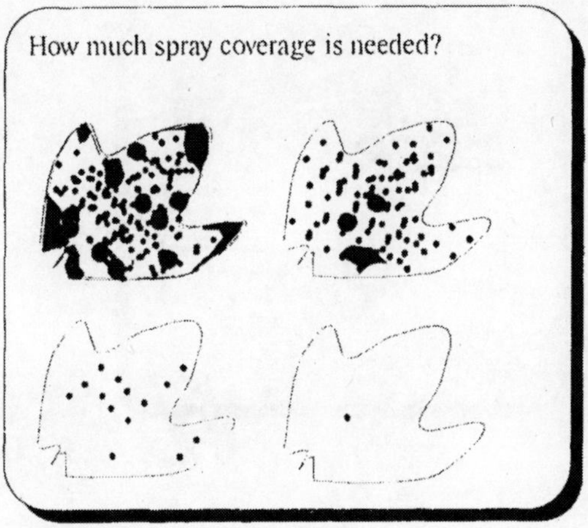
How much spray coverage is needed?

Later in the day as the sun warms the earth and soil and air temperatures rise, small droplets with a low fall velocity may move upwards against gravity in 'convection' air currents. Such upward currents of air are replaced by cooler air moving downwards. This air movement causes TURBULENCE.

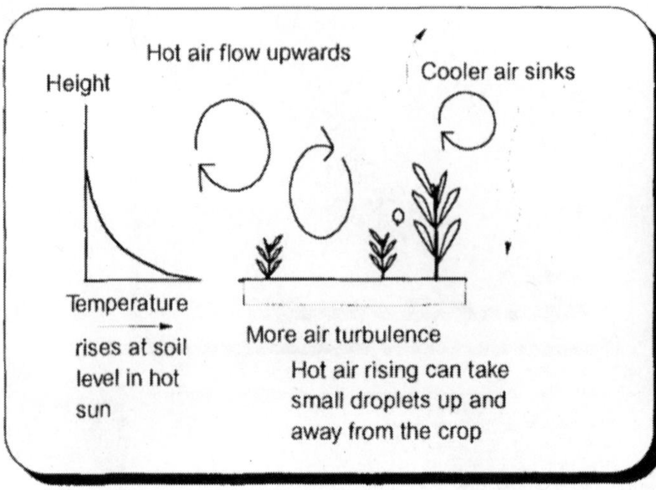

If there is sufficient wind across a field, then the air movement will be also disturbed by the 'friction' caused by the crop, so increasing TURBULENCE within the crop canopy. This turbulence and the extra force of the wind changing the velocity of the droplets will increase IMPACTION of the spray droplets on plants.

Very large droplets may bounce off leaves and the smallest may drift outside the treated field, so in selecting spray nozzles, the aim is to choose a spray droplet spectrum to transfer as much as possible of the pesticide to the biological target.

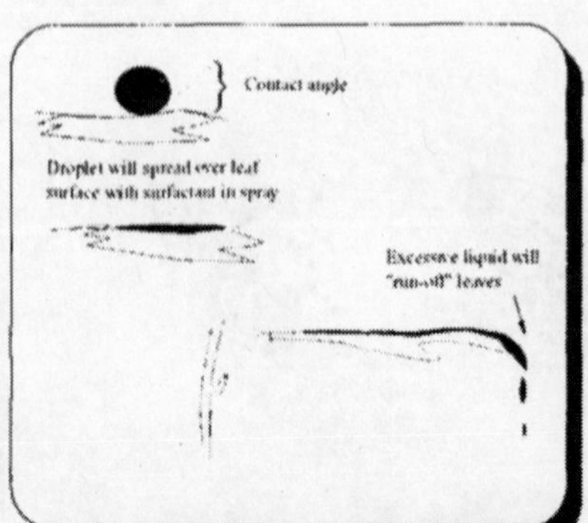

With hydraulic nozzles, which produce a range of droplet sizes, pressure control is considered very important and is discussed later. On some sprayers the size of droplets can be selected and this is considered in Chapter 6.

Rain

Having invested in purchasing a pesticide, the farmer will not want to see the chemical wasted by rain washing spray deposits off treated surfaces. Usually if two to three hours of dry weather follow a spray, it will have been effective, especially if the pesticide is absorbed into the plant or even in the waxy surface of leaves. In the tropics rain can occur within two hours, so under these conditions an adjuvant to increase penetration and rain fastness may be advisable. In some cases, the cost of the adjuvant is not justified, when a repeat spray is needed to treat new growth after a few days. In some cases light rain or dew can be beneficial in redistributing pesticide deposits.

Other Factors

Other factors which need to be considered in relation to pesticide application are as follows:

The growth stage of the crop - will it compensate for early damage? Is the crop at a stage that is susceptible to a particular pest?

The presence of natural enemies - will they control the pest?

The severity of damage and pest numbers - a spray may be too late if damage is already severe.

Are beneficial insects in the area - will bees be affected by a pesticide treatment?

When will the crop be harvested - will residues be on the harvested produce?

Many people think that spraying a pesticide is easy. On the contrary, when all the factors are carefully considered, the application of a pesticide is a complex process, which requires information and skills from different dis'ciplines. Application is NOT easy and requires training to ensure the correct pesticide has been applied at the right time as efficiently as possible with minimal disruption of the environment.

Crop Specific Application of Pesticides

The following specific examples of pesticide application on selected crops and pests are to illustrate how methods of using pesticides can be improved.

Cotton

Cotton crops are attractive to insect pests throughout most of the season. However, there are many cultivars that are resistant to some of the pests, thus in most of sub-Saharan Africa, plants with pubescent leaves are grown because they are resistant to jassid infestations. Where suitable varieties are sown, it is usually possible to avoid insecticide applications during the first 6 - 10 weeks, any damage being compensated for later in the season. During this period natural enemies are also conserved. Farmers are then advised to monitor their crops from the start of first flower buds, usually six weeks after germination, and if insect numbers exceed ACTION THRESHOLDS select an appropriate insecticide and apply the lowest effective dosage. The application according to an economic threshold is often difficult as the farmer may not know the income expected from a particular crop due to changes in market prices. Nevertheless there are pest populations that from previous experience indicate clearly that 'action' is needed.

In Central Africa a vertical 'tailboom' mounted on the back of a knapsack sprayer has been recommended since 1960. This allows the volume of spray and hence the dosage, with constant spray concentration, to increase by adding an extra pair of nozzles to the 'tailboom' for each 30 cm height of the plants, up to a maximum of 4 pairs of nozzles. The aim is to increase the volume application rate from 50 to 200 litres per hectare in proportion to the leaf area of plants. The volume was increased by 50 litres/hectare for each 30 cm increase in plant height by adding two nozzles to the tailboom. Insecticides were provided in sachets so that one sachet was used with each knapsack load.

The main difficulty with the knapsack spraying is that in the drier areas water supplies are poor, so reduced volume techniques were developed. The spinning disc sprayers are held with the spray applied downwind of the operator so in contrast to use of a knapsack sprayer with a lance in front of the operator, there is less risk of operator contamination. These sprayers can be used either to apply ultra-low volume oil-based formulations (1 -3 l/ha) or very-low water-based formulations (5 - 20 l/ha), preferably with some evaporation retardant in the spray. Molasses has been used in some areas, and similarly a refined mineral or vegetable oil, such as soya, cotton or rape seed oil, mixed with emulsifier is available as an adjuvant. Reduced volume techniques have been widely used in West Africa in francophone countries.

Figure 2 Cotton tailboom on Knapsack Sprayer.

In these countries if the spinning disc applicator is not used preference has been given to a rear-mounted horizontal boom to spray two rows, but the spray distribution of a downward directed spray is poor due to the umbrella effect of the upper leaves.

Many other techniques have been used on small-scale farms, including syringe type pumps and compression sprayers. The syringe pumps, frequently used in India, are laborious to use and the operator always gets heavily contaminated by spray. Compression and lever-operated knapsack sprayers with a simple lance also result in operator contamination, due to the operator walking into the sprayed crop canopy. In China, some growers have used a simple brush to paint stems of selected isolated plants infested with aphids. Using a systemic insecticide this localised treatment aims to reduce the pest infestation without disrupting natural enemies.

Rice

Many rice ecosystems have good populations of natural enemies to check insect pest *infestations. It is* crncial therefore *to ensure* that insecticides are not used unless the pests are causing major crop loss. Where rice is grown extensively for domestic consumption in contrast to the sale of cotton for cash, it is much more difficult to establish an action threshold. However in some countries rice is grown for export. FAO has been actively encouraging rice farmers to recognise beneficial insects and spiders that need to *be* conserved and avoid rushing *to* apply an insecticide when there are very few pests. Furthermore the application of broad-spectrum insecticides is discouraged.

There are some occasions when an insecticide may be needed, but unfortunately when farmers have decided to spray rice crops, they tend to swing a lance through 180° in front of them as they walk through the crop to treat a wide swath. The better way is to have a short boom carried downwind from the operator. Compared with cotton growing, there has been virtually no development of reduced volume spray application using spinning discs on rice.

These spray techniques are satisfactory for leaf feeding pests such as *Hispa,* but it is difficult to get penetration deep into the canopy if there are nymphs of the brown planthopper feeding at the base of the plants. Many growers have also used low concentration granules of a systemic insecticide, such as 3 per cent carbofuran, distributed by hand. The main concern with this method is the direct contamination of irrigation water. In Japan, motorised granule applicators (see Chapter 7) have been used.

Many rice growers are now changing from transplanted to direct seeded crops, and this has increased the application of herbicides. Although farmers are using a single lance, the distribution of the herbicide would be better if growers used a short side-mounted boom with fan nozzles, operated at constant pressure.

Coffee

Fungicides to control leaf rust ideally need to be applied to the undersurface of leaves. In some countries the terrain has allowed the use of tractor-mounted air-assisted sprayers or the invelied hockey-stick boom. However, in many countries coffee is grown on steep slopes and manually-carried equipment is essential.

In Colombia some trials were carried out using a rear-mounted vertical boom similar to that described for cotton. (See Chapter 4) The nozzles were angled upwards to direct spray between the layers of leaves. This approach will also improve the penetration of spray to the berries, needed for beny borer control. Until now most growers use a single lance, while some invest in motorised mistblowers. The operator therefore projects the spray forwards with the consequent risk of contamination and its distribution depends on how the operator directs the nozzle. Pruning the coffee bushes is needed to facilitate passage through the crop and assist spray coverage. A new type of air-assisted spinning disc sprayer has been designed to project spray behind the operator. It applies reduced volumes which offers the potential of higher work rates as well as increased operator safety.

Weed control in coffee is similar to many plantation crops, where weeds are most important in the establishment stage, and then competition needs to be minimised and access through the crop maintained. A knapsack sprayer is used frequently with a deflector nozzle to provide a single swath matching the row width or pathway through a crop. Evenness of spray deposits can be improved by having a constant pressure at the nozzle and training operators to walk at a uniform speed. Spinning disc sprayers have also been used. A slightly higher deposit on the edge of a pathway is useful, where weeds are encroaching into the path.

Maize

The main concern is to control weeds early in the season. Most farmers have used traditional hand hoeing, but there is a limit to the area that can be weeded, so that if a farmer spends too long weeding the cereal crop, other crops are sown late. The greatest problem is weeding betw~en the young plants; hoeing or pulling weeds by hand can disturb the roots. A herbicide treatment can be confined as a band to control weeds along the crop row, leaving the inter row to be cultivated. This requires an even-spray fan nozzle or a deflector nozzle preferably with a narrow angle to confine the herbicide to a narrow swath.

One major weed in some areas is the parasitic weed *Striga*. The weed is sometimes not noticed until it is too late to reduce its immediate effect on the crop. However, if it is left to seed, the problem will increase. Individual patches of *Striga* can be spot-treated before flowering with a hydraulic sprayer fitted with a cone nozzle. Alternatively if there is a large area of the ~eed, a deflector or fan nozzle can be used. A low pressure is needed to avoid drift of the herbicide to adjacent crops.

The early infestation of stalkborer can be selectively treated with a very small quantity of insecticide applied as a granule to the 'funnel' of maize plants. A simple pepperpot type applicator can be used and in some patls of Africa farmers have punched small holes in a plastic bottle to make an applicator.

Summary

Chemical control needs to be undertaken only as part of an integrated pest management programme. Careful selection of the least hazardous and most selective pesticides is needed to harmonize with cultural and biological controls. When a pesticide needs to be applied, the weather conditions need to be considered, especially in relation to time of application. In hot, dry sunny weather small spray droplets are liable to be carried upwards by thermal air movement. Some wind is needed to assist impaction of droplets on foliage, and by turbulence mix the spray cloud with air in the crop canopy. Some aspects of application on particular crops, such as cotton, are described.

CHAPTER2

HYDRAULIC NOZZLES

The pesticide application equipment most commonly used by small-scale farmers is the lever-operated knapsack or compression sprayer, both of which use HYDRAULIC NOZZLES.

As the selection of nozzle is crucial to good application details of the hydraulic nozzles are given before describing specific types of sprayer. Other types of nozzle are referred to in Chapters 5, 6 and 7.

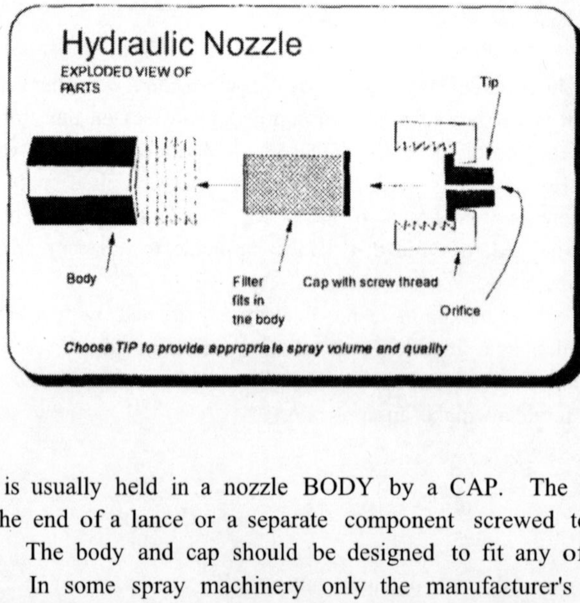

The nozzle TIP is usually held in a nozzle BODY by a CAP. The body can be an integral part of the end of a lance or a separate component screwed to the end of the lance or a boom. The body and cap should be designed to fit any of the nozzle tips described below. In some spray machinery only the manufacturer's own nozzle is supplied, but this should be discouraged so that users can fit the most appropriate nozzle tip for a particular application. The body may incorporate a diaphragm check valve that acts as a anti-drip device (see also Chapter 3 regarding Spray Management Valve.)

The cap usually screws onto the body, in which case some nozzle tips need to held by a spanner as the cap is tightened, so that the direction of the spray is correctly aligned. Some caps have a bayonet fit, and are particularly important when using fan nozzles on a boom. All the parts of a nozzle should fit together by hand without any leakage.

Diaphragm check valve

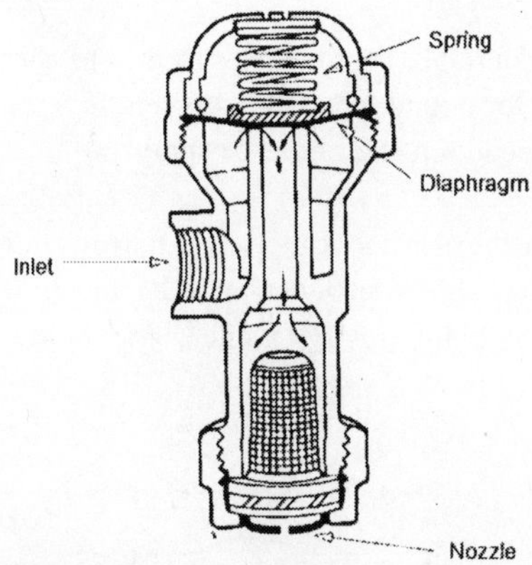

The nozzle tip is a small but essential part of these sprayers as it:-

a) meters the volume of spray liquid applied

b) projects the spray in a particular pattern eg fan or cone

c) breaks the liquid into spray DROPLETS.

The correct choice and operation of the nozzle ls of paramount importance for accurate application of any pesticide.

At the nozzle, the spray liquid is forced through a small hole or ORIFICE under pressure. The pressure provides the energy to form a sheet of liquid which then breaks up to form the spray. As the pressure is raised, there will be an increase in liquid flow, the angle at which the spray liquid is emitted is wider and the size of droplets is smaller. Ideally the pressure should be CONST ANT at the nozzle. Recommended pressures are:-

i) 1 BAR (100 kPa) for HERBICIDE sprays

ii) 3 BAR (300 kPa) for INSECTICIDE or FUNGICIDE sprays

iii) 2 BAR (200 kPa) may be used as a compromise between the two pressures for any pesticide.

Do not increase the pressure to increase the flow rate of a nozzle. Always change the nozzle tip to one with a larger orifice to provide a greater output. Alternatively where droplet size is crucial, use more nozzles.

Most nozzles are held about 50 centimetres from the surface being sprayed. This allows the spray sheet to form and produce droplets over a particular SWATH. The width of the swath will depend on the angle of spray.

Holding a nozzle closer to the surface can give a narrower swath, but if the distance is greater than 50 centimetres, the swath may not be much wider as the distribution of spray droplets will be affected by gravity, while some of the smaller spray droplets can be moved downwind.

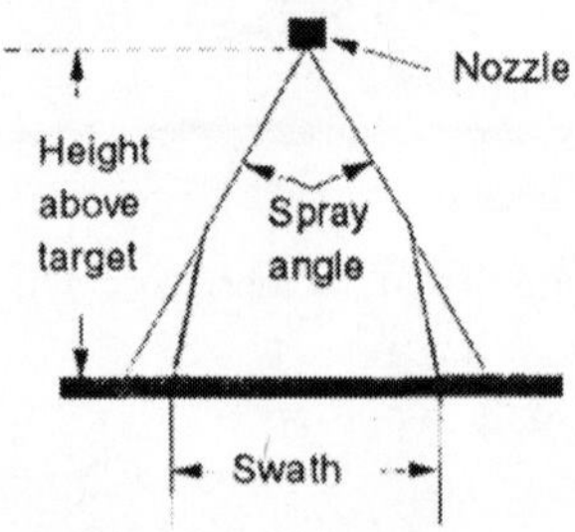

Note swath is narrower than theoretical as droplets are falling by gravity; wind will affect actual swath depending on droplet spectrum.

Table 2. The width of spray swath (cm) on surface for different spray angles and heights.

Nozzle Angle	Swath width for Height above surface	
	30cm	50cm
65°	38	63
80°	50	83
110°	85	142
130°	129	210

Types of Hydraulic Nozzle

Several types of hydraulic nozzle can be used. The choice of nozzle type will depend on the pesticide being applied and the surface (eg. crop, insect, weed, soil, structure, etc.) being treated.

Main Types of Nozzles

Fan pattern	*Cone pattern*
Standard fan	Hollow cone
Low pressure fan	Full cone
Pre-orifice fan	Variable cone
Even spray fan	
Deflector	

Each of the above and adaptations of them are described in the following section:

Deflector (D)

The deflector nozzle, sometimes called an impact-, anvil- or flooding nozzle, has a round opening through which liquid passes before hitting a flat surface that deflects the fan-shaped spray towards the target surface. More liquid is at the edges of the sheet, so the spray pattern of most deflector nozzles is not as uniform as with other fan nozzles; however, the circular orifice is less likely to block. Some new designs of deflector nozzles give a better spray pattern, suitable for use on a boom.

The deflector nozzle is ideal for spraying pathways with herbicides, such as in tea, coffee, oil-palm and rubber estates. Some have a wide angle so that the full width of the pathway can be covered in one swath.

Standard Fan (F)

The most widely used fan nozzle has an elliptical orifice, that results in more spray being emitted through the centre of the nozzle and less at the edges. These spray patterns of adjacent nozzles can be easily overlapped to provide a uniform distribution of spray, so they are used along a boom, particularly on tractor-mounted and other mechanised sprayers. They are used for applying pesticides to relatively flat surfaces, such as soil surface treatments. They are also used extensively to treat closely spaced cereal crops and are sometimes set at an angle to increase deposition on the upper foliage.

Off-set Fan (OC)

The orifice is set at an angle so that the spray is directed to one side of the operator. They are useful when the swath is to the side of a pathway. They have not been used as much of other types.

Double outlet Fan

These nozzles have two identical orifices, one is set to spray at 30° forwards while the second is angled 30° back. When applying the same volume as a single orifice fan nozzle, the double outlet nozzle produces a finer spray. They can be used as an alternative to cone nozzles in some circumstances as the distribution of spray across the swath is better, but the small narrow orifices are prone to blockages. They are used for pre-emergence herbicide application.

Low Pressure Fan (LP)

These nozzles are similar to the standard fan but were designed specifically for herbicide applied at 1 bar pressure. Flow rates are specified for 1 bar rather than 3 bar. They are used where there is concern that drift can occur.

Even-spray Fan (FE)

When the whole area does not require treatment, but the spray has to be uniformly sprayed across a band, the even-spray nozzle is recommended. In general, the spray droplets are smaller than if a standard fan is used at the same pressure and output. They are particularly used where a band treatment of herbicides along the crop row can be integrated with cultivation of the inter-row. They can also be used for a band treatment across the inter-row between crop rows.

Pre-Orifice Fan. (RD)

With increasing concern about spray droplets drifting away from the target surface there has oeen an increasing interest in reducing the proportion of small droplets produced by a hydraulic nozzle. The presence of an orifice before the final elliptical orifice results in a drop in pressure and on average the droplet size is larger than if a standard fan nozzle is used. These nozzles are marketed to produce less drift, but fewer small droplets can be a disadvantage for some pesticides.

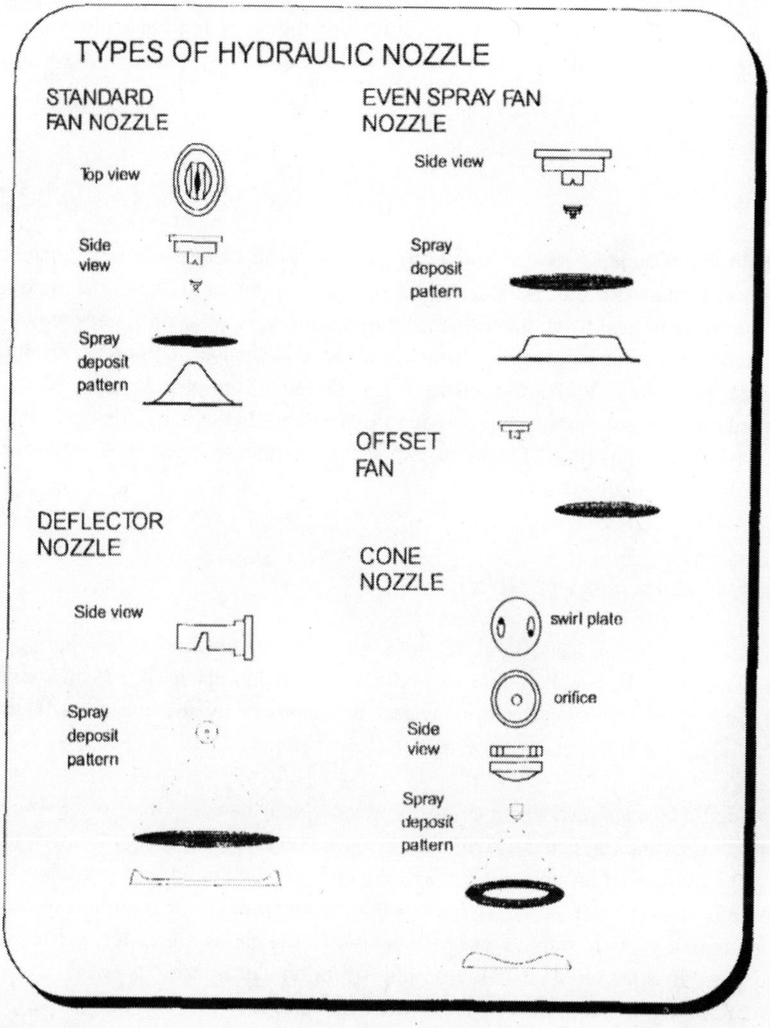

Hollow Cone (HC)

This nozzle has a round orifice in front of a swirl plate or core. Tangential slots in the swirl plate cause the liquid to flow in a circular manner through the orifice. This results in the emerging spray forming a cone, with very little spray in the centre; hence the name hollow cone. In some cone nozzles the swirl plate is integrated with the back of the orifice plate. As spray approaches surfaces from different angles, the cone nozzle is most appropriate for treating complex foliage, for example treating broad-leaved crops such as cotton with insecticides or coffee with fungicides.

Solid Cone

This is similar to the hollow cone nozzle, but some liquid is fed centrally through the orifice to provide a solid cone. They are not used extensively and are most suited to spot treatments.

Variable Cone

In this nozzle the distance between the swirl plate and the orifice can be adjusted. The closer the swirl plate to the orifice, the wider the spray angle and the smaller the droplets that are produced; at the other extreme, spray is in a very narrow jet. One type of variable cone has frequently been fitted to the cheaper compression sprayers, but to adjust the spray angle, the operator must touch the nozzle tip. This is not recommended for safety reasons. Also it is extremely difficult to adjust this type of nozzle to a particular position. On more expensive equipment there is a hand-operated trigger to adjust the angle of spray.

Materials of Construction of Nou.les

Typically most nozzle tips used in agriculture are now manufactured in a hard-wearing engineering plastic. 'Kematal' is one example of the materials used. If an unsuitable plastic is used in nozzle manufacture, water is absorbed by the plastic affecting the flow rate by changes in the size of the orifice.

Plastic-moulded nozzles are cheaper to produce than nozzles that were previously made from brass, and in general will spray effectively for as long as brass nozzles. However the lifetime of any nozzle tip will depend on the liquid and particles that are forced through it under pressure. Where water is obtained from a stream or borehole and is contaminated with minute abrasive particles, such as sand, the orifice will be eroded rapidly. It is essential that the output of individual nozzles is checked regularly (see CALIBRATION) and a tip replaced when necessary.

More expensive tips can be purchased, and these are made from stainless steel and ceramics, but the extra cost is only justified where extensive use of nozzles is expected with abrasive particles in the spray.

Colour Code

The use of plastic nozzle tips allows manufacturers to colour-code the tips to indicate the flow-rate, litres per minute, at a set pressure. The colour code does not indicate the type or angle, ONLY THE OUTPUT.

The international colour code is as follows:

Colour	*litres/mimte at 3 bar pressure*
Orange	0.4
Green	0.6
Yellow	0.8
Blue	1.2
Red	1.6
Brown	2.0
Grey	2.4
White	3.2

Nozzle Code

Each manufacturer has a paliicular method of marking individual nozzles. For example 8001 and O1 - F80 are the equivalent nozzles from two manufacturers. The manufacturer may include a one or two-letter code to indicate the type of tip.

However there is a code developed by the British Crop Protection Council to describe a nozzle, independently of commercial interests. This code consists of four parts - the type, angle, output and operating pressure.

The nozzle types are:

D	deflector
F	standard fan
FE	even-spray fan
RD	pre-orifice nozzles (reduced drift)
LP	low pressure fan
HC	hollow cone
OC	off-set fan

The spray angles most commonly used are 80°, 110° or 65°, but other angles are available. In particular with hollow cone nozzles, it is possible to put together various combinations of disc and swirl plate to achieve particular spray angles.

The output is given in litres per minute.

The operating pressure is usually in BAR. Note 1 bar is equivalent to 100 kilopascals (kPa), lkgfper centimetre squared or 14.7 pounds per square inch (psi).

Typical examples of the nozzle code are as follows:

FS0/0.4/3 is an 80° standard fan nozzle, emitting 0.4 litres per minute at 3 bar pressure. It is identical to the manufacturer's code example, 8001, given above.

Fl10/1.2/3 is a 110° standard fan nozzle emitting 1.2 1/min at 3 bar.

D /0.46/1 is a deflector nozzle, 0.46 1/min at 1 bar (Note: angle is not specified here)

Spray Quality

Nozzle manufacturers indicate in their catalogue the quality of spray obtained with the different nozzle tips and operating pressures. These quality assessments are based on measurements of the droplet spectrum. The droplet spectrum will affect the extent of spray coverage and the extent to which some of the droplets will be carried by the wind. In general a FINE spray is needed when insecticides and fungicides are applied to achieve good coverage of the target leaves or buds. A spray of MEDIUM or COARSE quality, with more of the volume in larger droplets, is used with herbicides to reduce the risk of downwind DRIFT on to adjacent unsprayed areas. If no information is given on the pesticide product label, always use a MEDIUM quality spray for most agricultural applications.

Determination of the spray quality is by measuring the spray droplets in flight using laser particle size analysers. The actual measurements will depend on which of the different laser systems is used to produce the data, so it is important to compare each nozzle with data from a set of reference nozzles. The spray quality for a particular nozzle can change according to the pressure; the higher the pressure, the finer the spray.

An indication of the size of droplets in the spray is given in the following table:

Type of spray	*Volume Median Diameter*
Aerosol - fine	<25 µm
- coarse	25 - 50 µm
Mist	50 - 100 µm
FINE spray	100 - 200 µm
MEDIUM spray	200 - 300 µm
COARSE spray	300 - 400µm
VERY COARSE	>400 µm

Sprays with a VMD < lOOµm may be referred to as very fine sprays, when used in agricultural or outdoor horticultural situations. Indoor applications of aerosols in the United Kingdom are generally referred to as fogs, whether produced by a thermal or cold fogger.

Aerosols are generally not used in agriculture, except to treat horticultural crops inside glasshouses and stored produce in warehouses (see Chapter 7). In open fields the droplets of aerosols ar so small they are blown too easily by the wind. Mists can be used in some air-assisted treatments of trees and with some ULV and VLV sprays containing sufficient involatile liquid to prevent the droplets becoming too small by evaporation (see chapters 5 and 6).

Filtration

The small orifice in a nozzle tip can easily be blocked by small debris, so all users should filter the spray at different stages of the application system. There should be a filter inserted in the nozzle body. The size of the mesh should be smaller than the size of the nozzle orifice. A 50 mesh filter is generally recommended for most manually operated sprayers. A finer mesh is very likely to block frequently, while a coarser filter is unlikely to prevent nozzle blockages except for very high output nozzles.

There are several types of filter, but the type usually referred to as a 'thimble' has a larger surface area and is suitable for most applications.

Filter Mesh Sizes

Fine

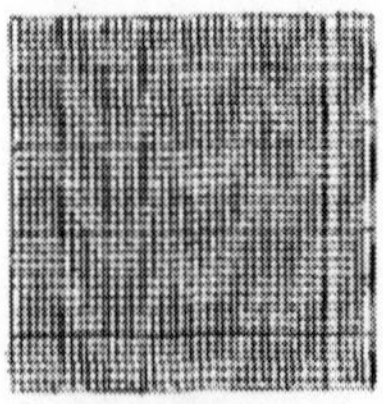

or Coarse

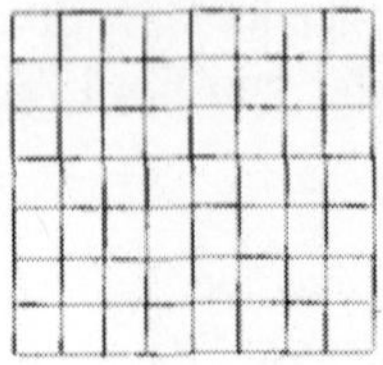

20 mesh/inch	0.8mm aperture
30 mesh/inch	0.6mm aperture
50 mesh/inch	0.3mm aperture
80 mesh/inch	0.18mm aperture
100 mesh/inch	0.14mm aperture

Nozzle output litre per min.	Filter before pump	Filter between pump & outlet valve	Filter before nozzle
< 0.8	50	100 or 80	100
0.8 - 3.0	30	30 or 50	50
>3.0	20	20 or 30	30

Calibration

The user needs to know the volume of spray applied to a particular area of crop so that the correct amount of pesticide is added to each tank-load of spray.

Where the user can do the calculations the following technique can be used:

Method A

1. Put clean water into sprayer, check for leaks.

2. Spray water into a bucket for ONE M[NUTE and measure the volume. A small graduated container is supplied with some sprayers.

3. Measure the swath width by spraying water on a dry surface.

4. Measure the distance you can walk though your crop in ONE MINUTE, while spraying.

The calculation is as follows:

$$\frac{\text{Volume per minute (litres per minute)}}{\text{Swath (metres) x Speed ofwalking (metres/minute)}} = \text{Litres / minute}^2$$

This answer can be multiplied by 10,000 to give litres/hectare.

As an example

$$\frac{\text{0.6 litres per minute}}{\text{0.5 x 60 metres/minute}} \text{ x 10000} = \text{200 litres/hectare}$$

Therefore if each tankload is 15 litres

200/15 = 13.3 loads per hectare.

If you need one litre of the pesticide product per hectare, then 1000ml divided by 13.3 = 75ml per load.

Method B

This eliminates some of the calculation by using a calibrated bottle, known as a 'KALIBOTTLE' to collect the spray while treating 25 metres^2. The bottle is graduated in litres per hectare. The user needs to measure a distance of 25 metres if the swath is 1 metre wide, or adjust the distance in relation to the swath width so that the same area is treated. Thus if the swath is 0.5 metres, then measure out 50 metres. The nozzle tip is removed and the cap of the 'Kalibottle' attached to the end of the lance. The nozzle tip is replaced, and the bottle screwed into its cap. The operator then sprays the measured area in the normal way and reads the volume of liquid collected on the bottle.

Method C

Put a small volume of water into the sprayer tank and operate pump to check for leaks and that the nozzle is operating correctly. Then when all the liquid has been pumped from the tank, put in a known volume - say 1.5 litres into a 15 litre tank. Spray part of the crop and measure the number of rows treated. This number multiplied by 10 will give the number of rows treated by one tank load. From this the number of tankloads needed to cover the whole of the crop area can be estimated. If 12 tankloads are needed for one hectare, then the dosage per hectare divided by 12 equals the amount that has to be added to each tankload.

Figure 3 Kalibottle used to calibrate Sprayer

USEFUL EQUATIONS

Calculating Nozzle Flow

$$\text{Required Volume (l/ha)} \times \text{Nozzle spacing (cm)} \times \text{Speed (km/h)} \div 60{,}000 = \text{Flow through one nozzle (l/min)}$$

Calculating Speed

$$\text{Flow through one nozzle (l/min)} \times 600 \div \text{Required Volume (l/ha)} \div \text{Nozzle spacing (cm)} = \text{Speed (km/h)}$$

Calculating Litres per Hectare

$$\text{Flow through one nozzle (l/min)} \times 600 \div \text{Nozzle spacing (cm)} \div \text{Speed (km/h)} = \text{Required Volume (l/ha)}$$

Note:- Ideally PRESSURE is fixed in relation to spray quality required.

Choice of Nozzle

In relation to manual carriage of water, it is preferable to keep the volume per minute as low as possible to avoid constantly refilling the sprayer. However, a smaller orifice is more likely to block even with an appropriate filter in the nozzle body. Deflector nozzles are used mainly for herbicides applied with knapsack sprayers, when only a single nozzle is fitted to the lance. Standard fan nozzles are used for herbicide applications usually where several nozzles are on a horizontal boom. Cone nozzles are used mainly for insecticides and fungicides where the treated surface is not uniform or
flat, for example spraying leaves of a crop like, coffee, cotton or potatoes.

				Standard fan	Deflector	Cone
Herbicides	Pre-Emergence		Overall	Recommended	Recommended	Not recommended
			Band	Use even spray		
	Post-emergence	Contact	Overall	Recommended	Recommended	Can be used against certain weeds
			Band	Use even spray		
		Systemic	Overall		Recommended	
			Band	Use even spray		
Fungicides			Contact	Cereals, rice, wheat, etc.		Recommended for most foliar targets
			Systemic			
Insecticides			Contact			
			Systemic			

The usefulness of a particular nozzle can be checked by attaching small pieces of water sensitive card to the important areas of foliage, such as the undersurface of leaves, where pests are present. After spraying with water the coverage on individual cards can be readily seen by the blue spots where droplets have impacted on the yellow card. As the card is so sensitive to water, the foliage must be dry and the user must have dry fingers. Water sensitive cards can be obtained through nozzle suppliers. If the special card is not available, a glossy white card, such as Kromekote card used by printers for the covers of magazines, may be used if the spray liquid contains a colour dye. Some operators add a dye to their spray to check coverage.

Nozzle Selection

FINER ←→		COARSER
Hollow cone	TYPE	Deflector / Whirl
Smaller	SIZE	Larger
Larger eg 110°	ANGLE	Smaller eg 65°
Higher	PRESSURE	Lower
Greater hazard	DRIFT	Less hazard

The stain of droplets on paper surfaces will always be larger than the actual droplets, as they flatten on impact with the surface. Thus if the size of stains is measured, the data has to be corrected for the spread factor, which will vary with different formulations of spray and the surface characteristics of the card.

As some pesticide formulations can leave a white deposit, spray deposits with a crop canopy will show up on pieces of black card which can be fixed to leaves. Table-tennis balls can be painted with blackboard paint and pinned to cotton plants.

Summary

Most small sprayers have been provided with only one nozzle, but there is a very wide range of nozzle tips that are available. As water supplies are often limiting, the lower output nozzles should be selected, but the orifice should not be too small, otherwise it may block easily if proper filters are not used. A small range of different tips should be selected for the different pesticides and crops within a farming area, for example a deflector nozzle for general weed control, an even-spray fan nozzle for band treatment and a cone nozzle for applying insecticides and fungicides. With all nozzles calibration is essential.

CHECKING that the RIGHT NOZZLE Is being used Is ESSENTIAL with any equipment.

CHAPTER3

LEVER-OPERA TED KNAPSACK SPRAYERS

The most widely used small-scale sprayer in the developing countries is the lever-operated knapsack which requires constant pumping while· spraying. A tank, usually of 15 litre capacity, is carried on the shoulders using two adjustable straps. In some countries a smaller 10 litre tank is preferred to reduce the weight and be more appropriate for physical stature of the farmers. A third waist strap is recommended to ensure the tank fits tightly to the operators back to ensure that the effort of pumping is efficiently transferred to the pump. The lever operating the pump is usually under the arm and should be easily adjusted for either the left or right arm. Underarm sprayers are generally considered to be more comfortable than those in which the lever is operated over the shoul.der. However, in some crops, movement of the arm well clear of the crop is advantageous. In rice and cotton many farmers prefer the over-arm lever.

The lever is connected to either a piston or diaphragm pump. The pump may be mounted external to the tank for ease of maintenance, but is more frequently inside the tank where it is protected from damage if the sprayer falls from the operator's back.

Piston pump

These are preferred for applying insecticides and fungicides at pressures of 2 or 3 bar (200 - 300 kPa), or higher. They are also better where several nozzles are mounted on a boom. Spray liquid from the tank passes a valve and is drawn into the pump chamber by an upward movement of the piston. The reverse stroke closes the first valve and forces liquid past a second valve into an air chamber. This air chamber forms part of the piston on many sprayers. As ·liquid is forced into the air chamber, pressure increases, and is used to direct the liquid through a dip tube and hose to the trigger valve and lance. A paddle attached to the internal pump mechanism provides agitation of the spray liquid. In a few sprayers, hydraulic agitation is provided. Modern designs use a synthetic plastic elastomer as a seal for the piston. A recent design has twin pistons so that spray is delivered on both up and down strokes of the lever.

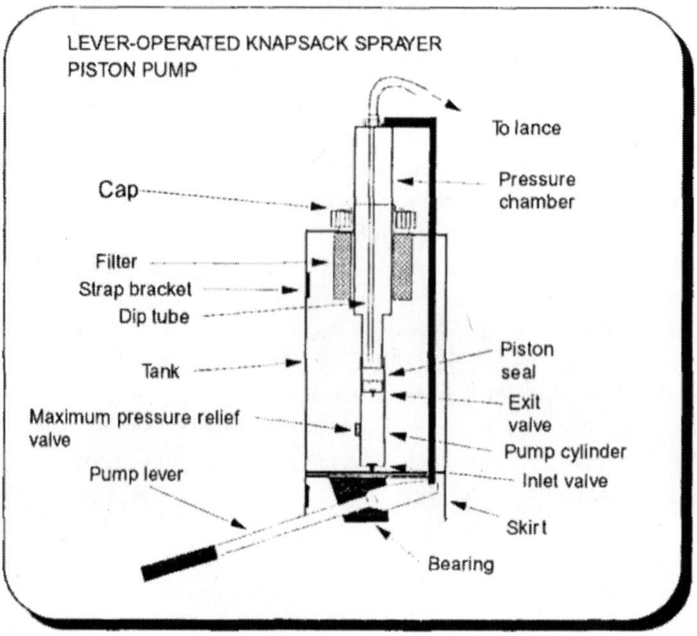

Diaphragm pump

Instead of a tight sealed piston, this pump has a flexible elastomer diaphragm, usually mounted in the base of the tank. Movement of the diaphragm up and down is much less than with a piston, and normally these pumps are more suited to low pressure operation (1 bar; 1OOkPa). However, with a single low output nozzle, it is possible to achieve pressures up to 3 bar (300kPa). There is usually less wear on the pump with abrasive particles compared with the piston pump, but the diaphragm must be able to withstand chemical degradation due to some formulations.

Air Chamber.

This is an essential part of all these sprayers as the cushion of air pressure evens out the extreme fluctuations in pressure with each stroke of the pump. On many of the cheaper sprayers the capacity of the air chamber is too small and pressure changes still occur with each pump stroke. Therefore the capacity of the chamber should be at least ten times the output displacement of the pump. If water fills the chamber reducing the volume of air, the pressure fluctuations will also occur.

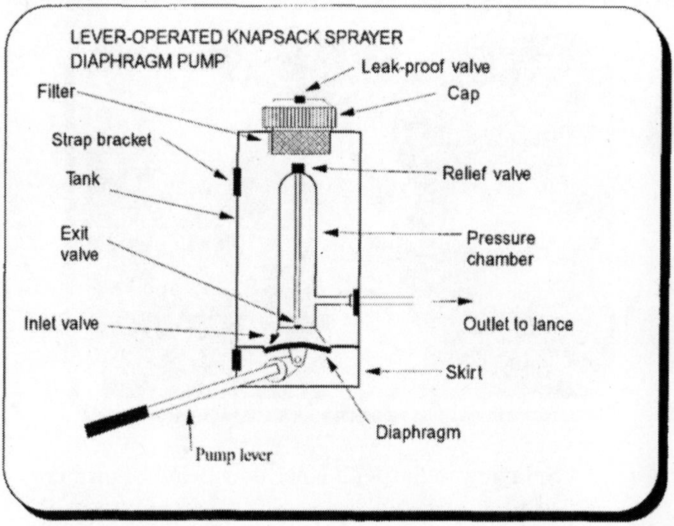

Despite the air chamber, the pressure of spray at the nozzle is very dependent on the way the operator is pumping. Full strokes of the pump are needed to ensure efficient I use of the valves in the pump. As the pressure will still vary at the nozzle, ideally a pressure control valve should be used.

Some manufacturers have fitted a spring-loaded valve in the air chamber. These could be adjusted or changed for different pressure settings, but as this involved inserting a hand inside the tank, contaminated with pesticide, these valves should now be regarded as a maximum pressure safety valve. When they open, the pressure tends to drop rapidly, so they are not designed to provide a constant pressure. (see also Spray Management Valve).

Tank opening

The tank opening needs to be large (usually > 10 cm) to allow rapid filling of the tank, from a bucket. A large capacity filter should be inset in this opening. The design of this filter is important to avoid restriction of water flow, avoid splashing of the operator and also remove any debris which could damage the pump or valves. The tank opening has to be sealed by a lid which can be easily removed, even with gloved hands. On some sprayers there is an air vent in the lid, but this must be protected by a valve to prevent liquid leakage.

Figure 4 Spraying cotton with lance downwind of operator

Figure 5 Knapsack with rear-mounted horizontal boom.

Trigger valve

The operator can control the period of spraying with a manually-operated trigger valve. On most sprayers the handle of the valve houses a filter to avoid the valve mechanism being blocked by small pieces of debris which may escape earlier filtration. The valve is usually fitted with a lock to keep the valve open for long periods of application. However, there should also be a lock to hold it closed to avoid any accidental discharge of spray. The trigger valve should be comfortable to hold for prolonged periods, with smooth surfaces to avoid accumulation of any spray deposits. Some valves are thumb operated, but these are more tiring to use.

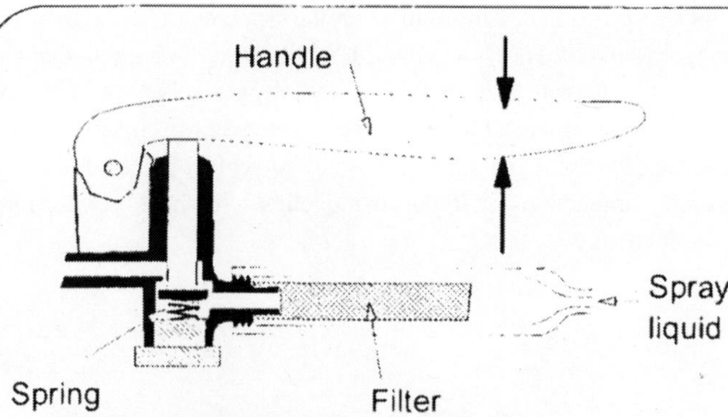

Trigger control valve
The valve should have a catch to prevent accidental opening. A catch is often provided to keep valve open for long periods of spraying.

Figure 6 Spraying herbicide after sowing maize.

Hose

Spray is transferred from the air chamber to a trigger valve by a flexible hose, usually made of a reinforced elastomer material resistant to chemicals. Some tubing has a pesticide resistant inner lining with a more flexible outer casing. A poor quality hose may soften under high tropical temperatures and should not be used. The length of the hose will depend on the position of the outlet on the sprayer relative to a comfortable position for the trigger valve. The hose should not be too long as it can catch in the crop. The connections between the hose and the trigger-valve at one end and the sprayer should be held by a replaceable clip. The worm-drive clip is considered most suitable as they can be reused if the hosing has to be replaced. Some manufacturers have designed reusable connections without using clips. All hose connections need to be checked regularly for leaks.

Lance

The lance should be at least over 50 cm in length and preferably longer to emit spray well away from the operator to minimise contamination. In some countries farmers are advised that the lance must be 100cm long. The end of the lance may be straight, but most have an angled section referred to as a swan-neck, which allows spray to be directed downwards. The connecting threads on the lance should enable extra sections to be added. Some manufacturers can provide an extendible lance, for example when spraying trees. Other attachments are available to protect trees while spraying herbicides.

The thread at the end of the lance should also accommodate a standard nozzle body so that any of the nozzle tips and associated filter described in Chapter 2 can be used. A short boom can be added to the end of a lance to use several nozzles, but when more than one nozzle is used, the total output must be within the capacity of the pump. Ideally the pump should be capable of delivering 90 litres per hour at 3 bar pressure with 30 strokes of the lever per minute.

Nozzle/position

Unfortunately most users of lever-operated knapsack sprayers hold the trigger valve and lance in front of their body, often waving it from side to side, and walk directly into the spray or treated foliage. This inevitably results in the operator being heavily contaminated by the pesticide spray. Ideally the lance should be held downwind of the ?perator.

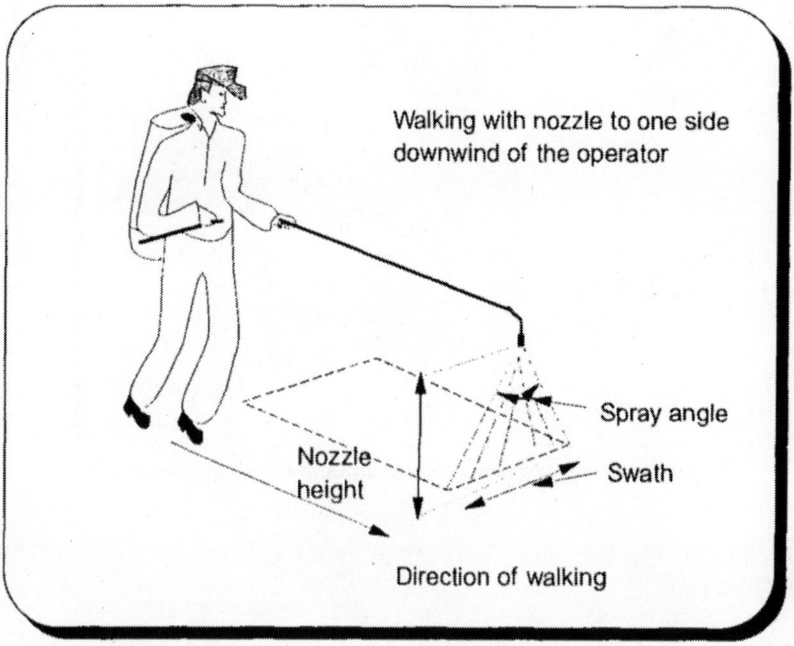
Walking with nozzle to one side
downwind of the operator
Spray angle
Nozzle
height
Swath
Direction of walking

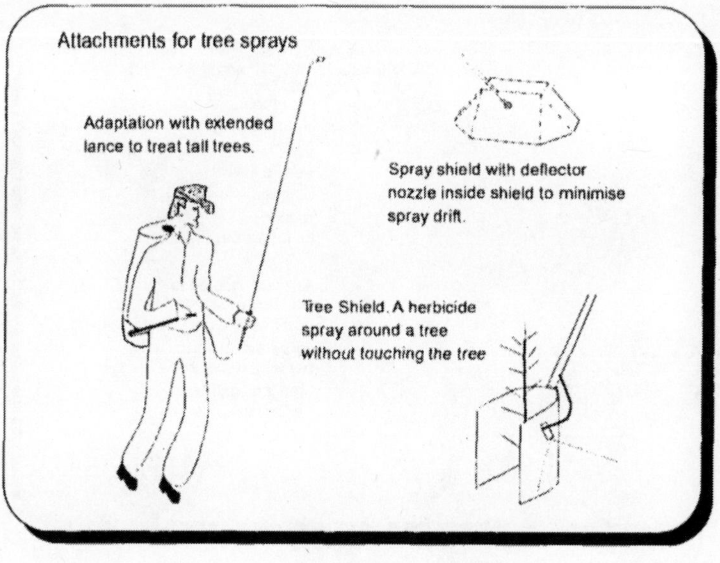
Attachments for tree sprays
Adaptation with extended
lance to treat tall trees.
Spray shield with deflector
nozzle inside shield to minimise
spray drift.
Tree Shield. A herbicide
spray around a tree
without touching the tree

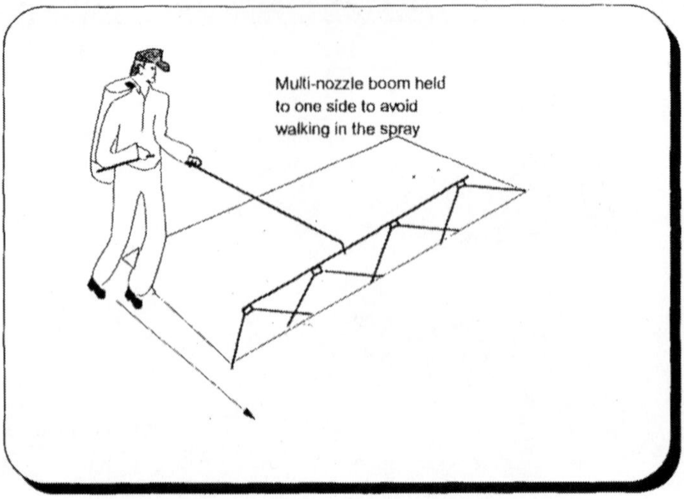

Offset fan nozzles are available to treat a swath to the right or left of the operator, but these are rarely used.

Several attempts have been made to design booms which are mounted on the sprayer tank and carried to the rear of the operator. A horizontal boom was used in West Africa to treat cotton, while in Central Africa, a vertical 'tailboom' was introduced to improve spray distribution through the crop canopy, as well as reduce operator contamination.

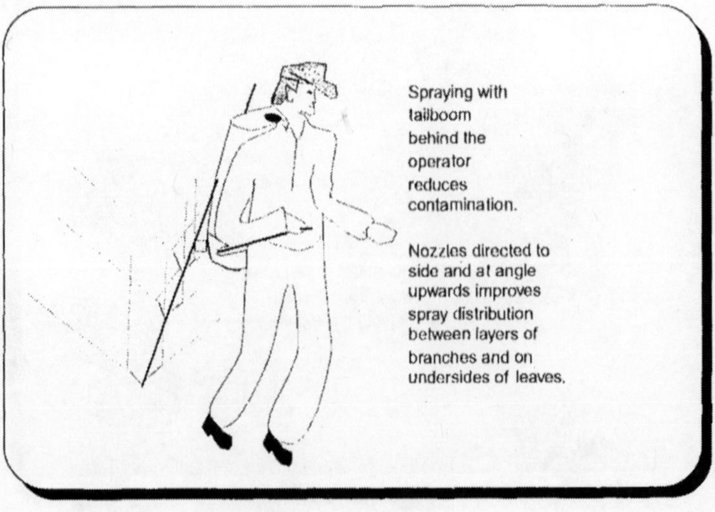

The cone nozzles (60°; 196 ml/minute output) were also directed upwards at an angle to project the spray between the layers of leaves and ensure deposition on the branches and undersurfaces of leaves. This contrasts to many situations where users spray down on to leaves, which then act as umbrellas and prevent deposition on plant surfaces most exposed to pest infestation.

Even with a simple lance, it is often better to spray upwards though foliage. A twin nozzle holder with swivel mounting can be used to angle the nozzles and are sometimes used to direct spray to either side of the inter-row.

Spray Management Valve (SMV)

This valve referred to earlier provides a uniform pressure at the nozzle as well as providing a non-drip positive shut-off This avoids the trickle of pesticide that is emitted after the trigger valve has been closed. It consists of a diaphragm check valve which opens at a preset pressure. In front of the diaphragm is a piston, the shape of which restricts the flow of spray liquid to the nozzle, when the pressure exceeds the set pressure. The valve can be fixed to the trigger valve, or closer to the nozzle at the end of the lance.

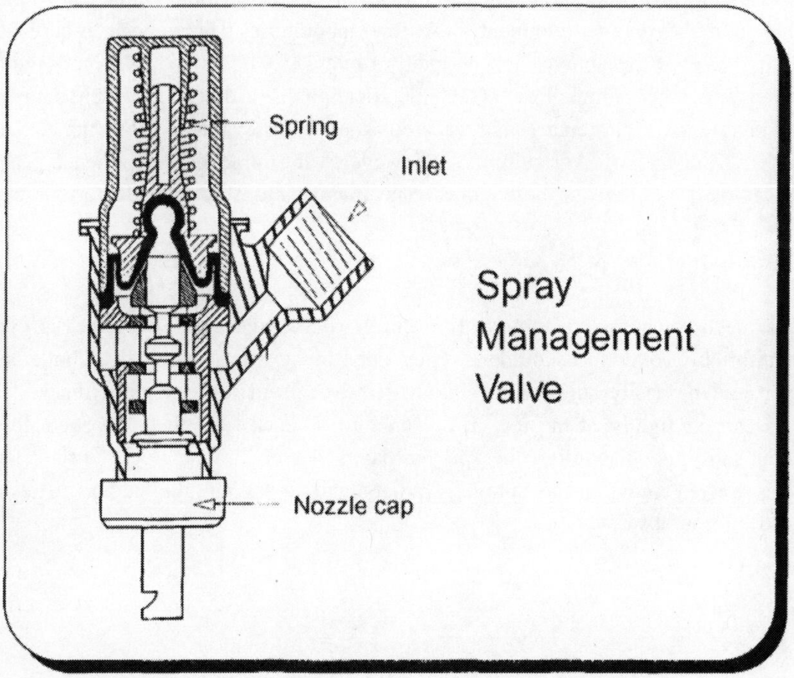

The colour coded valves are:

yellow	1 bar (lOOk:Pa)
blue	2 bar (200kPa)
red	3 bar (JOOkPa)

The use of a spray management valve is particularly beneficial on manually operated sprayers as it relieves the operator of the responsibility of maintaining the correct spray pressure. Without the SMV a user of a compression sprayer is likely to start spraying with a high pressure that falls during spraying, but with the SMV pressure is constant, thus avoiding a change in application rate with time. With the lever-operated sprayers the fluctuations that occur with different pumping strokes and the variation between the way in which individual operators use the pump lever is eliminated.

Choice of Knapsack

Many users purchase the sprayer which is readily available where they are farming, or which is the cheapest. These may be the sprayers manufacture locally. Often the cheapest sprayers are not very durable, frequently leak pesticide and do not apply chemicals very accurately. Nevertheless well-designed modern lightweight sprayers are now available at low cost. Some countries have specifications aimed at improving the quality and safety of equipment. A few laboratories have been established to simulate field use of the sprayer for periods of over 250 hours to assess the durability of the pump and other components (see Specifications). Advice on which sprayer can be most suitable in a particular area or crop should be available from the Extension staff of the Ministry of Agriculture. However, the practical experience of local successful farmers is often the best guide. The main features to consider are:-

Straps

The straps need to be wide enough (>5 cm.) so that the weight of the sprayers is spread comfortably over the shoulder. They need to be firmly attached at both ends to the sprayer and be easily adjustable without removal from the operator's back, so that the tank is carried tightly at the top of the operator's back, and then loosened for easy removal for refilling. Too often the sprayer hangs loosely on the back so the effort of pumping is wasted moving the sprayer, rather than operating the pump. The straps should also be non-absorbent.

Weight

The total weight of a full sprayer should be less than 25kg to avoid back strain of the operator. Where possible with large scale spray programmes, a table or platform placed at the mixing site will eliminate the need to raise a full sprayer from ground level. The weight of sprayers with plastic components is generally less than metal sprayers.

Filling

Modern designs of sprayer have a large opening to make it easier to fill the sprayer with an open bucket. However, where possible ancillary equipment should be provided to assist with correct mixing. In one country a special bucket with a cover over half the top surface and a filter in a large spout was made locally for cotton farmers to help measure the knapsack load and pour it in the tank without spillage.

Leakage

A careful check should always be made with clean water, even with a new sprayer, to ensure it has been assembled correctly and does not leak from any part in any position and at the operating pressure. A major cause of operator contamination is from leaks.

Pump

The operator should put the sprayer on and adjust the straps correctly and then check that the position of the pump lever allows a full stroke to be made easily and at a steady rate.

Nozzles

It is essential that the operator can change the type of nozzle or replace a blocked nozzle easily. On many of the cheaper sprayers only one nozzle is provided, and this may not be suitable for certain pesticides. Some locally made nozzles are of inferior quality and do not give an acceptable result. (see also Chapter 2)

Tank

The design of the sprayer should allow complete emptying so that no pesticide spray remains either in the tank or pump after the recommended cleaning has been carried out.

Durability

The overall construction of the sprayer should be sufficiently robust that with routine care and maintenance, it will continue to operate efficiently for a minimum of three seasons. The sprayer should require the minimum of maintenance, which is easy and requires few tools. The local supplier should keep spare paiis in stock and additional nozzles. The sprayer should be provided with an adequately illustrated instruction manual that gives safety advice, spare parts list and operating instructions.

Motorised Hydraulic Knapsack Sprayers

These sprayers have a two-stoke engine to drive a pump, but unlike the motorised mistblowers described in Chapter 5 there is no fan unit. The equipment is considerably more expensive than a lever-operated knapsack sprayer, but their use avoids the manual effort of pumping. They are usually fitted with a double acting diaphragm pump. There is a lance and trigger valve as on the manual sprayers but the pressure can be pre-set and maintained.

More recently a sprayer with an electrically driven pump has been introduced with power from a rechargeable battery.

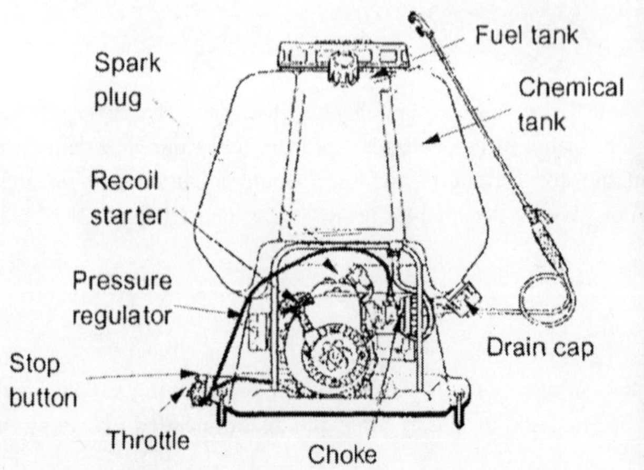

Motorized hydraulic knapsack sprayer

Summary

The lever-operated knapsack sprayer will continue to be the most widely used sprayer on small-scale farmers and where the terrain prevents using mechanised equipment. Users should select durable, easy to use sprayers, check that it works well without leaks and choose appropriate nozzle arrangement to minimise operator contamination. Before use the sprayer should be calibrated and after spraying washed before storage.

CHAPTER4

COMPRESSION SPRAYERS

The simplest and cheapest small sprayer suitable for farmers with a very small area of crops that need protection is the compression sprayer. Some are very small with a capacity of less than 5 litres and are usually hand-carried. For the small-scale farmer the compression sprayer with a capacity of I Olitres is recommended. While many of these were fabricated from metal, the trend is to make plastic sprayers. They consist of a tank, with an air pump, usually fitted through the filler hole of the tank, and a delivery hose with trigger valve, lance and nozzle. Other details of the lance are included in the following chapter as they are similar to those used on lever-operated knapsack sprayers.

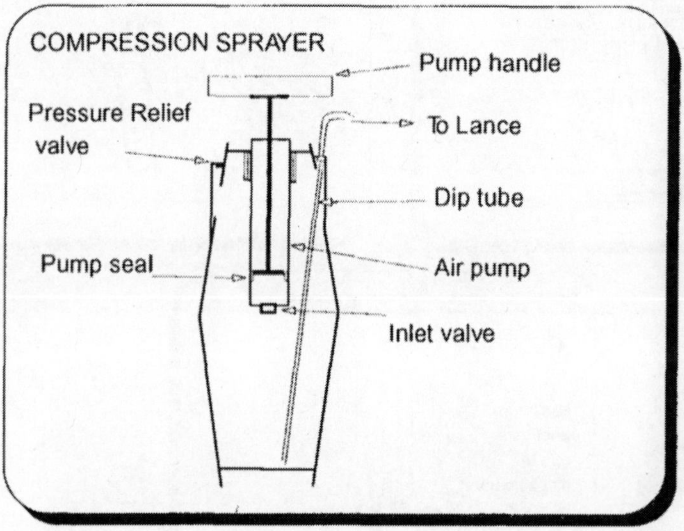

When using these sprayers, the spray liquid is poured into the tank, making sure that the maximum filling capacity is not exceeded, usually about 75 per cent of the total capacity of the container. For a small sprayer where the tank is not too heavy to lift and shake, it is preferable to put water in first, then add the pesticide, shake the tank for preliminary mixing. The remaining quantity of water can then be added. Contamination of another container to prepare the spray is avoided in this way. However some formulations may need to be pre-mixed before pouring into a sprayer.

The pump is replaced in the tank opening and fitted tightly to avoid air pressure escaping. The operator then pumps air into the container. As the cheaper versions of this equipment are not fitted with a pressure gauge, the operator has to judge when to stop pumping, by the feel of the effort needed to use the pump. A pressure relief valve should be fitted to prevent over-pressurization.

The operator does not need to keep pumping while spraying, but may need to repressurise the tank by more pumping to discharge all the spray liquid from the tank. The air pressure is used to force the liquid along a delivery hose to the lance, trigger valve and nozzle. Initially the pressure is high, but as liquid is sprayed, the pressure falls, affecting the performance of the nozzle.

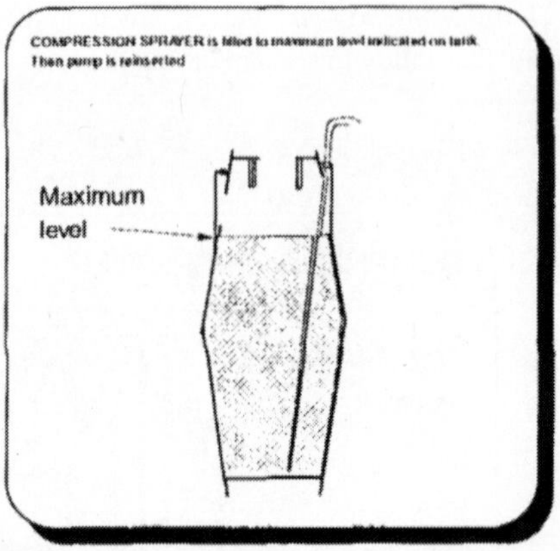

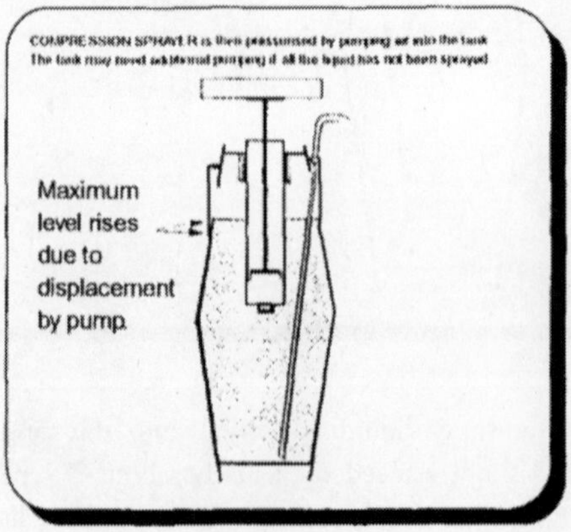

Some manufacturers have fitted an adjustable pressure control valve to ensure the pressure is constant at the nozzle. Experience with these is that they are relatively expensive, so seldom are fitted as standard, and, if supplied, they are frequently adjusted incorrectly. In preference a Spray Management Valve (see previous Chapter) is now recommended, as this has a preset pressure which cannot be adjusted by the user in the field.

The main advantage of a compression sprayer is that the user can concentrate on directing the nozzle while spraying as pumping is intermittent. This is particularly important where wall surfaces or stacks of grain are being treated in stores.

When all the liquid has been sprayed, some air pressure may remain in the tank. Equipment is designed so that as soon as the tank lid/pump is loosened by unscrewing it, air will escape, making a hissing sound. This hissing should be allowed to stop before opening the tank to refill it.

Most of these sprayers are used with a single low-output nozzle to avoid refilling the tank too frequently. If a nozzle with an output of 0.5 litres per minute is used, a ten litre sprayer has to be refilled after 20 minutes. The lance should be at least 1 metre in length to deliver spray away from the operator's body. The threads on the lance should allow extension lances to be added so that when used in stores, the spray can be released to treat the roof area.

Figure 7 Compression sprayer with rear-mounted boom for spraying coffee.

Some compression sprayers are fitted with a pump which is separate from the lid. These were developed primarily for vector control and have been used to spray surfaces on which mosquitoes may rest. They are quite suitable for agricultural use, but tend to be heavier and more expensive.

Other pressure-retaining types, which are still used in a few tropical countries, for example in Colombia to treat coffee, are pre-pressurized with air and then the spray liquid is pumped in to the tank. Due to the much higher pressures used with these sprayers, the tank is much heavier and a pressure control valve is essential. Their use should be phased out as the pressures used are no longer acceptable and the weight is also considered excessive for manually-carried equipment.

Handymist sprayer

In China a unique type of compression has been developed in which air is continually pumped into the tank, partly to maintain the air pressure on the spray liquid, but also to deliver some of the air separately to the single nozzle. The air atomizer nozzle produces a very fine spray.

Syringe type sprayers.

The simplest spray applicator is a manually operated syringe pump. Some have a double-acting pump to give a more continuous flow of liquid. In India they can be purchased with a plastic tank carried on the operator's back, but they are very tedious to use and the amount of spray applied and operating pressure will vary with each stroke of the pump. As the nozzle is at the end of the syringe it is difficult to direct the spray while pumping. In addition they are not recommended, because the operator is always more heavily contaminated.

There are also some small shoulder-mounted sprayers which require constant pumping, but allow the operator to direct a lance and nozzle more easily. They are relatively inexpensive, but the spray delivery is poor due to variations in pressure and the low capacity of the pumps.

Summary

When only a small area needs treatment infrequently, choose a compression sprayer with a pressure regulator. Use a low output nozzle and hold it downwind while spraying.

CHAPTERS

MOTORIZED KNAPSACK MISTBLOWERS

Conventional hydraulic sprayers can only project spray droplets generally over a short distance from the nozzle. Thus to project spray upwards into trees or across several rows of crops, it is necessary to add an airstream to carry droplets towards the target. The manually carried sprayer with a fan to create the airstream is generally known as a motorised knapsack mistblower. They are designed to produce a very fine spray or 'mist' (50 - lOOμm VMD droplets) and apply lower volumes than conventional knapsack sprayers (eg 20 l/ha rather than 200 l/ha), but most machines can also be adapted to apply granules and dusts (see Chapter 7).

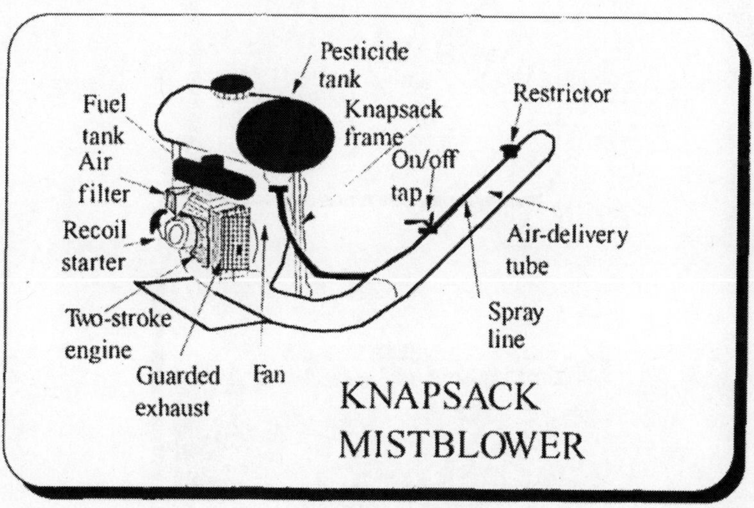

These sprayers consist of a 35 - 70 cc two-stroke engine which drives a centrifugal fan. The larger size engine is required to drive a fan with a greater output of air volume. These heavier sprayers are needed to spray taller trees as the greater volume of air emitted can project droplets higher than the small mistblowers. It is rarely possible to project droplets higher than lOm vetiically even with the larger motorised knapsacks.

These engines need specialist maintenance, so their large-scale use should be restricted to areas with qualified mechanics who are able to service the equipment. In extensive areas of small farms, mobile workshops are an effective way of assisting users who would otherwise have difficulty in transporting their equipment to a central workshop.

The engine and fan unit are attached by anti-vibration mountings to a knapsack frame, designed to allow the sprayer to stand upright on the ground. The frame, with straps, also carries a pesticide tank, spray deliver tube, fuel tank and an air delivery hose. A nozzle is mounted at the end of the air delivery tube. The volume of spray liquid emitted is controlled by a variable or fixed restrictor, and there is an on/off tap also attached to the air delivery tube.

The tank is usually of 10 litre capacity. Some with larger tanks are made, but the extra weight in addition to the fan and engine, is considered unacceptable. The tank has a wide opening to facilitate pouring liquid into it. The floor of the tank should also slope to a low outlet point.

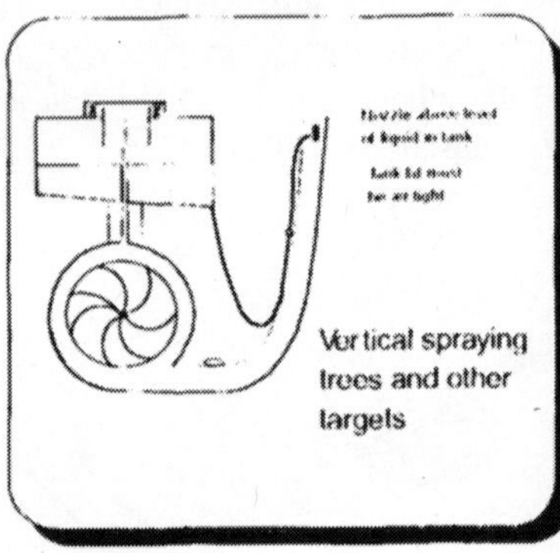

Figure 8 Motorised Knapsack Mistblower.

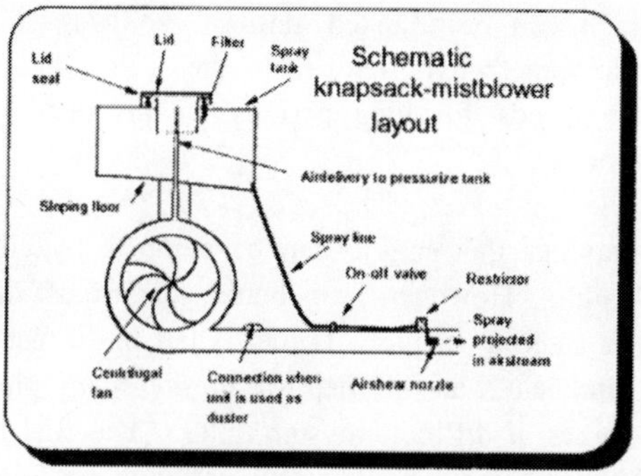

Some air is fed from the fan into the spray tank and usually ducted to the base of the filler hole filter to provide low pressure (0.25 bar; 25 kPa) to deliver the spray liquid to the nozzle. This air pressure is most impoliant if the standard air delivery tube is pointed upwards, when the nozzle may be above the level of liquid in the tank. The large lid on the tank must therefore have an air-tight fit. On some machines, instead of relyirg on this air pressure, there is a separate pump, especially if the air delivery tube is extended to treat very tall trees. This pump is usually mounted directly on the same drive shaft as the fan, but may be driven by a belt.

Engine

The two-stroke engine is operated with a fuel containing standard petrol and oil mixed in a particular proportion. The mixture, often I pali of oil to 25 parts of petrol should be clearly indicated, preferably by an embossed marking on the fuel tank. SAE 30 oil may be used, but manufacturers advise avoiding any multigrade oil, as the additives in these products will adversely affect the life of the engine.

When starting the engine, the CHOKE on the CARBURETTOR is closed to enrich the fuel mixture delivered to the engine. A float valve on some carburettors may need to be depressed to assist flow of the fuel to the engine. With the THROTTLE closed, the engine is then started by pulling the recoil starter rope sharply and evenly. The rope should be allowed to rewind slowly and not to snap back, nor should it be pulled to its full extent. The rope/handle may need several pulls before the engine starts. If it fails to start, the spark plug may need to be removed and cleaned. When the engine has started, the throttle can be opened up slightly, then the choke is opened. When the engine is then running, its speed can be increased using the throttle. If the recoil starter is missing or broken, it is possible to use a strong piece of rope to start the engine. These engines should always be used at their full speed to ensure proper lubrication of the moving parts. At the proper speed the fan draws sufficient air over the engine to cool it. The speed can be checked by using a tachometer such as a 'Vibratak'. This has a wire, the length of which can be adjusted until it oscillates at its maximum. The speed of the engine and fan can then be read from the scale. The engine should never be allowed to idle at low speeds for long periods as this results in overheating and carbon build-up in the exhaust.

On the completion of spraying, the engine can be stopped by closing the throttle or short-circuiting the spark plug. However, it is better to turn off the fuel tap and allow the engine to run until it is starved of fuel. If fuel is left in the carburettor and engine, the petrol will evaporate and leave an oily deposit on the spark plug and block the fuel metering orifices. This makes it difficult to start later. The fuel tank and carburettor should be drained of fuel, if the equipment is not going to be used again within three days. This is particularly important in the tropics where high temperatures iricrease the evaporation of the volatile component of the fuel.

Always make sure the air filter is clean and the engine exhaust is covered by a guard. Avoid touching the guard as this can also get very hot after the engine has run for some minutes. As the noise level will exceed 90 decibels, the operator should wear ear protection, in addition to any other personnel protective clothing recommended.

Fan

The centrifugal fan is designed to produce a high velocity air jet (60 - 80 m/s) that on most machines is used to shear the spray liquid into droplets. If the nozzle is held too close to the crop, this high velocity of air may damage foliage and spray will be blown off leaf surfaces. This initial air velocity rapidly diminishes within the first 2m from the air delivery tube. The aim should be to hold the nozzle at least 2 metres from foliage.

Nozzle

The simplest of mistblowers have a single tube to direct the spray liquid into the high velocity airstream. However, several manufacturers have developed alternative ways of spreading the liquid thinly into the airstream. This may be an aerodynamic wing section across the airflow, or a circular disc. On some machines, the disc can rotate and there are special rotaly nozzles that can be fitted to most mistblowers; these can provide a more uniform spray droplet size distribution, but the quality of these devices does vary considerably. Flow rate and air velocity can also have a major effect on droplet size.

On these sprayers the flow rate is not determined by the nozzle, but by a restrictor mounted in line with the nozzle. On many sprayers, there is a variable restrictor, often with four settings. Users will frequently set this restrictor to the maximum open setting to empty the tank as quickly as possible. This may lead to poor atomisation, so the recommendation is to use a sprayer with separate fixed restrictors. When the appropriate restrictor is in place, it cannot be changed in the field.

On most machines it is possible to fit a very low output restrictor for spraying at ultra-low volume (see next chapter), but the droplet spectrum should be checked and a suitable spinning disc nozzle fitted, if necessary, to avoid wastage of spray in droplets which are too large. When ULV adaptors are used a filter should be fitted to the tank outlet to prevent the small metering orifice being blocked.

Calibration

With these machines it is not possible to collect the spray liquid discharged from the nozzle. It is usual therefore to put a measured quantity into a clean dry tank and measure the time required to spray all of it. The period of spraying should be at least one minute. The swath width will depend to some extent on how the operator uses the airjet. When spraying across a row crop, it can be a number of rows between successive paths across a field. In the case of tree crops, spraying may be confined to a single row and the volume per tree calculated on the basis of the time needed to project spray to all sides of a tree. Sufficient time should be given for the volume of air in each tree to be replaced with the air carrying spray droplets.

Examples:

A. Flow rate 0.5 litres per minute;
Swath 2 metres;
Walking speed 60 metres per minute.

Walking speed should be measured accurately by measuring the time to walk a measured distance, usually at least 100m at the speed the user would walk while spraymg.

Then the application rate is 41.67 litres per hectare.

B. Flow rate 0.24 litres per minute
Time required to treat individual tree is 0.5 minute
i.e 0.12 litres per tree; if 1000 trees per hectare,
then volume per hectare = 120 litres.

These sprayers are best used on tree crops, but are also widely used in some countries to treat bushes and row crops where penetration of the crop canopy is needed. In selecting this equipment users must accept the equipment is much more expensive and will require more maintenance.

See also information in Chapter 7 on the conversion of knapsack mistblowers for application of granules and dusts.

Summary

Where droplets have to be projected upwards into a tree canopy, the sprayer should provide an airstream, thus the manually carried motorised knapsack sprayer is needed, despite a higher capital cost and being more expensive to operate and maintain.

CHAPTER 6

SPINNING DISC APPLICATORS

The spinning disc type of sprayer is quite different from the manually operated sprayers described in previous chapters. They have a small plastic disc rotated by a motor, usually powered by batteries. They were developed to reduce the need to collect large quantities of water needed with hydraulic sprayers. In some developing countries, particularly in the semi-arid tropics, few farmers had adopted the use of knapsack sprayers to treat cotton and other crops, despite efforts by the Extension Departments to demonstrate large increases in yield due to spraying. The perception of the farmer is that it takes too long to collect and transport several hundreds of litres of water to the fields. Even the development of more boreholes to improve water supplies has little effect on the spraying of crops, because the water still has to be carried over a long distance. Many others seek water for other domestic uses, limiting the availability of the boreholes.

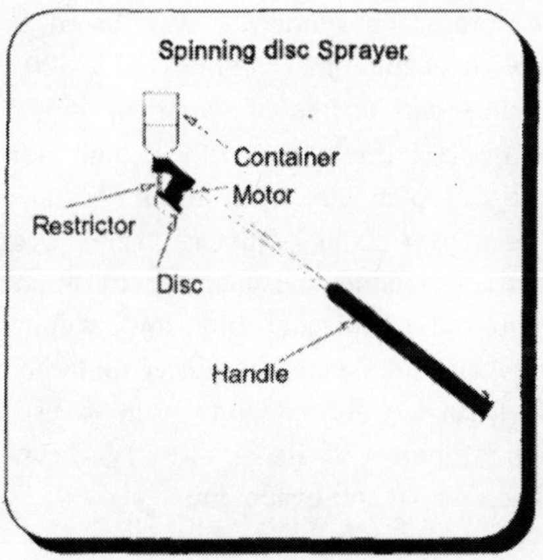

In consequence,yields remained low (<400 kg/ha seed cotton), until farmers accepted the use of spinning disc sprayers which, for example, are now used to treat cotton on over a million hectares of small farms in West Africa. These sprayers have also been used to apply insecticides or fungicides to groundnuts, pigeon pea, sorghum, millet, tomatoes and tobacco, as well as being the standard method for migrant pest control (i.e. locusts, grasshoppers and armyworrn) where the need to treat large areas rapidly means use of high application volumes is logistically unacceptable. By using smaller droplets which move downwind for ultra- or very-low volume insecticide and fungicide application, very high work rates are possible (typically 1 hectare per hour) using incremental spraying technique. However spinning disc sprayers can also be set to produce larger spray droplets for a placement technique if required, such as when applying herbicides.

Ultra-low volume

This is the minimum volume that can be applied and achieve economic control. This is usually less than 5 litres per hectare and because of the cost of special relatively involatile formulations, containing oil as a diluent, most ULV applications have been at less than 3 litres per hectare. Specific formulations, designated as UL formulations are available, which are not diluted by the farmer, thus helping to reduce calibration errors. When supplied in prepacked containers to attach directly to the sprayer they offer safety advantages by eliminating the mixing and loading procedure needed for other manually operated sprayers.

Verylow volume

As an alternative to ULV sprays, conventional water-based insecticide and fungicide formulations have been applied at minimum volumes of 5 - 20 litres per hectare. When spray volumes below 10 litres per hectare are applied, it is recommended to add an involatile adjuvant or choose a formulation with a high non-volatile component to restrict the extent to which droplet size can shrink in transport to the target. In Zimbabwe, molasses has been used as an adjuvant, that is also attractive to the pests. If 12.5% of the spray liquid is involatile, the diameter of droplets will decrease by half. Very-low volume sprays are now replacing ultra-low volume applications on many cotton farms in West Africa and are being introduced into other countries and crops. Conventional water-based herbicide formulations can be used through spinning disc sprayers at total application volumes of I0-50 litres per hectare, using large droplets which do not require the addition of anti-evaporants.

Controlled droplet application.

When using ultra- or very-low volume application, it is very important to avoid spray being wasted in droplets that are not effectively deposited on the target, particularly very large droplets that are likely to deposit on the soil rather than a crop or weed. Note that a large proportion of the spray volume with hydraulic nozzles can be very large droplets. The aim in foliar insecticide and fungicide application is therefore to produce a narrow droplet spectrum (50-1 SOμm diameter droplets) that will be carried by air currents into the crop canopy. Conversely, if the target is weeds, then the droplets should be larger (200 - 300 μm diameter) to avoid downwind displacement of the spray. Controlled droplet application (CDA) aims therefore to produce only the most appropriate band of droplet size to optimise deposition on the intended target. Droplet size is less important for application to soils, but much depends on the surface profile with clods or debris which intercept spray droplets. Nevertheless low volume CDA sprays can be used.

Figure 9 Spinning disc Sprayer.

Spinning disc Sprayers

Hand-carried, battery-operated spinning disc sprayers for CDA are generally of two types: sprayers intended for insecticide and fungicide application with small droplets (<150μm VMD), which will move downwind from their point of release, and others that produce large droplets (>200μm) suitable for placement, particularly herbicide application. The latter being equipped with a governed motor or equivalent system to provide a disc speed of no faster than 2000 rpm.

Each type of sprayer consists of a small plastic spinning disc rotated by a DC electric motor, powered by batteries. One manufacturer has used a spinning brush. The speed of rotation of the disc is usually dependent on the voltage of the power supply, so disc speed slows down as the batteries are used. Some sprayers have a battery case that can be adjusted to hold different numbers of batteries to change the input voltage. However, on some sprayers designed for herbicide application, a governor controls disc speed irrespective of the input voltage above a minimum voltage. A container of pesticide is attached so that liquid will feed by gravity through a restrictor to the surface of the disc.

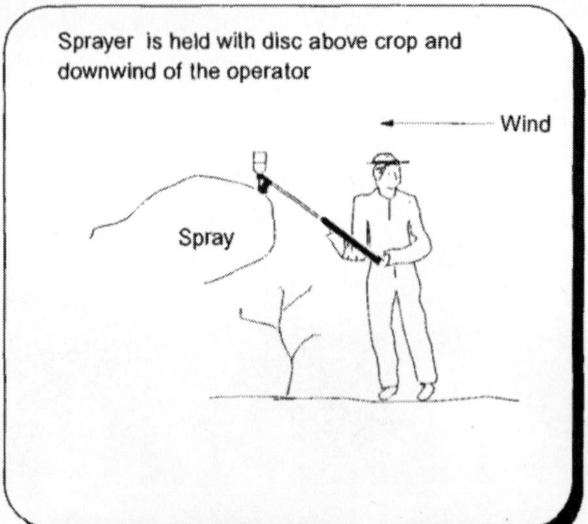

Flow control

The restrictors are generally colour-coded but there is no universal colour-coding standard at present. The range of restrictors is to allow the volume applied to the disc to be adjusted in relation to the viscosity of the formulation and optimise the droplet size obtained. The range of restrictors supplied is not to increase the volume applied per hectare but to maintain the optimum flow per minute in relation to the viscosity of the spray to obtain the appropriate droplet size. It is important to check the flow rate with the actual spray liquid under local conditions as viscosity can also be affected by temperature.

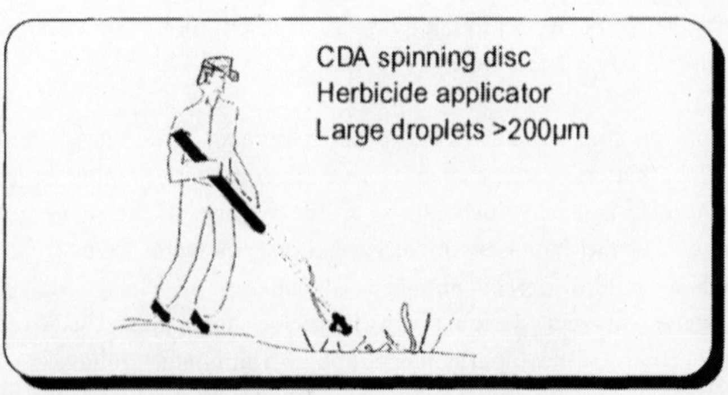

Disc design

On the earliest spinning disc sprayers widely used in Africa, a double disc was fitted, but subsequent research has shown that a single disc is better for the hand-carried equipment. A small single disc requires less power and in consequence, battery life is extended. The distribution of liquid evenly to the edge of the disc is important, so the better designs allow the liquid to spread into grooves, which individually feed a series of teeth at the edge. Control of flow rate is crucial to optimise droplet size, as single droplets are produced at very low flow rates, whereas as higher flow rates, the liquid leaves the disc as ligaments. The teeth are very important since the disc edge is where these spray ligaments or droplets are actually formed. Unless the flow is optimised there is less control of droplet size, especially if the disc is fed too much liquid.

Power supply

The main source of energy for these sprayers is the standard heavy duty 'D' size zinc-carbon battery. When using these dry batteries, their life is extended if the user avoids spraying continuously for more than 20 minutes, or no more than 2 hours on any one day with the same set of batteries. However, these are not always readily available, and in some countries, their quality has been poor. Alkaline manganese batteries have also been used, but the extra cost is only justified, if the sprayer is used daily over a prolonged period. In consequence, various attempts have been made to use rechargeable batteries. Lead acid motor cycle batteries have been used mainly on large estates where the problems of routine recharging can be managed. Solar panels to recharge nickel cadmium batteries have not been very widely used due to the capital investment required and difficulties in keeping batteries routinely charged. However future use could be a viable option as the use of solar panels for other purposes increases and their overall cost is more acceptable.

Power consumption is greatest for insecticide application at high disc speeds and if the volume of liquid is increased. For herbicide application with slow disc speeds, power consumption is low. Disc speed can be checked with a vibrating reed type tachometer, such as a 'Vibratak' in the field. Optical tachometers are available for laboratory use.

To avoid using batteries, one sprayer, known as the 'Birky' was designed so that the disc was rotated by air pressure, using a lever-operated air pump. On this machine the low disc speed was most appropriate for herbicide application. Other manufacturers are developing hydraulically driven discs, to be fitted to a knapsack sprayer, but these will again be primarily for herbicide application.

Track spacing

Using an incremental technique, the spinning disc sprayer, applying insecticides or fungicides, is held above the crop (usually about 0.5 m) and downwind of the operator. The distance between successive passes through the crop, the track spacing, will depend on the type and height of crop and amount of its foliage, as well as the wind conditions at the time of spraying. As wind speed is often very variable and can drop to zero, even with gusts of wind in excess of Sm/s, it is usual to avoid having too wide a track spacing so that parts of a crop are not left untreated. On small crops, the track spacing can be several metres, thus in Central Africa ULV sprays were recommended with a track spacing of 6m with cotton plants smaller than 30cm tall. This can be reduced to 4m when plants were between 30 and 60cm high and then to 2m for cotton plants over 60cm high. Other countries have tended to have a fixed track spacing regardless of the amount of foliage filtering the spray droplets.

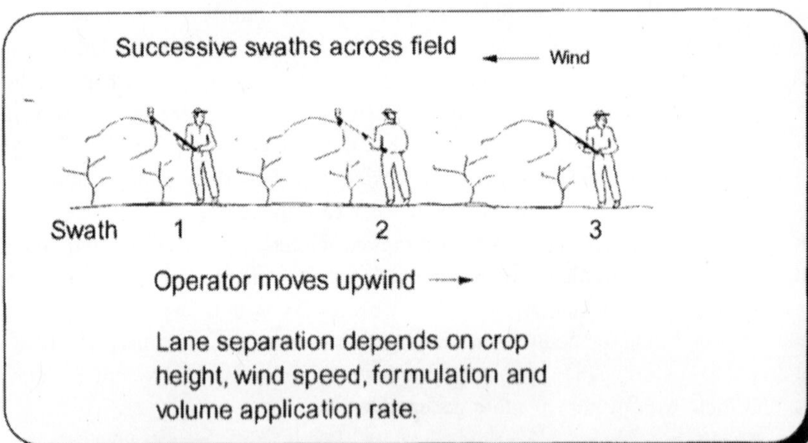

Height of disc above crop is usually 40 - 80cm.

Lane separation for Very-low volume may be 2 metres with low crops, but is reduced to 1 metre with plants over 50cm high.

Lane separation for Ultra-low volume oil-based sprays may be as wide as 6 metres for low crops, but is reduced progressively with plant height.

A sprayer with two spinning discs mounted on a vertical boom carried behind the operator has been recommended to treat tobacco with fungicides.

Using a placement technique to apply herbicides, with the disc usually held close to the ground, the track spacing is dependent on the disc diameter and speed of rotation. On herbicide sprayers with a disc diameter of 8cm., a rotational speed of 2000 revs/minute and a droplet size of about 250µm diameter, the swath is around 1.2m wide. They have been used to treat the circle around oil palms and along pathways in S.E.Asia. On some sprayers the disc may be partially shrouded, so that narrower swaths can be obtained. Liquid not emitted through the opening in the shroud is recirculated to the disc. There are also unshrouded narrow swath band applicators which are simpler to construct. The swath width for any particular herbicide applicator can be determined, while standing still, by treating a bare dry piece of ground, until the area wetted can be seen and measured.

Calibration

With the latest design of spinning disc, with openings on its surface, it is possible to allow spray liquid to be fed by gravity on to the disc and then collect the liquid draining frorri it, without rotating the disc. On earlier designs where the rotation of the disc affected the flow rate, the time for a known volume of liquid to be sprayed was measured. The swath width used in calibration is the track spacing selected. The actual swath width will vary with wind speed, droplet size and height of release of the spray. Walking speed is often approximately 60m/minute (1 metre per second), but it can be slower if passage through a crop is difficult.
The volume applied per hectare is calculated using the same formula illustrated earlier.

For example, with an ultra-low volume application, a flow rate of 30ml/minute, track spacing of 4m, and walking speed of 60m/minute, the spray is applied at:

$$\frac{0.030 \times J0{,}000}{4 \times 60} = 1.25\ \text{I/ha}$$

Similarly for a very-low volume application, a flow rate of 60ml/minute, a track spacing of I metre and walking speed of 60m/minute, the spray is applied at:

$$\frac{0.060 \times l0{,}000}{1 \times 60} = 10\ \text{litres/hectare}$$

With VLV application an EC, WP, SC or other formulation diluted in water may be used. If the application rate is 101/ha, the quantity of Product required per hectare is mixed in 10 litres of water. Thus if the sprayer container holds one litre, add one tenth of the quantity of Product per hectare to each 'full' container attached to the sprayer.

ALWAYS READ the LABEL.

Air-assisted Spinning Disc Sprayer

As there is no wind inside stores, glasshouses and polytunnels, a spinning disc sprayer fitted with a small propeller fan and powered by a rechargeable battery, mains supply or two-stoke engine can be used. The mist droplets will sediment on to horizontal surfaces fairly rapidly so the technique is not strictly a space treatment compared with the aerosol and fog treatments referred to in the next Chapter. Compared with a fog or aerosol treatment, the mist spray will allow localised treatment in a situation where there are some plants with a heavy pest infestation, but the grower needs to avoid killing natural enemies elsewhere in the glasshouse.

Spinning disc sprayers provide a narrower droplet spectra suited to a particular target; either small droplets for insecticides and fungicides, or larger droplets to avoid spray drift. They are used to apply reduced volumes of spray, mostly on cotton and other row crops in semi-arid areas by treating successive swaths as the operator moves upwind across the fields.

Figure 10 Spinning disc with fan behind operator.

Summary

Where water is limiting, especially in semi-arid areas, ultra-low volume oil-based and very-low volume water-based sprays can be applied with small hand-carried spinning disc sprayers. The size of droplets can be set by adjusting the speed of the disc, thus smaller droplets 50-1 SOμm are used for insecticides and fungicides and larger >250μm droplets for herbicides. Some motorised equipment is also available with a disc behind the operator to apply sprays to bush crops such as coffee.

CHAPTER 7

MISCELLANEOUS EQUIPMENT

Weed wiper

Several designs of weed wiper have been used to rub a translocated herbicide on the leaves of weeds. The simplest hand-held units consist of a plastic tube carried as the handle, from which the herbicide liquid is fed by gravity through a restrictor or compression joint to an absorbent material. The absorbent wick can be a piece of rope or sponge. The aim is to have the surface sufficiently wet that the herbicide is easily rubbed on to the weed leaves, but not to drip elsewhere, when damage to a crop may occur.

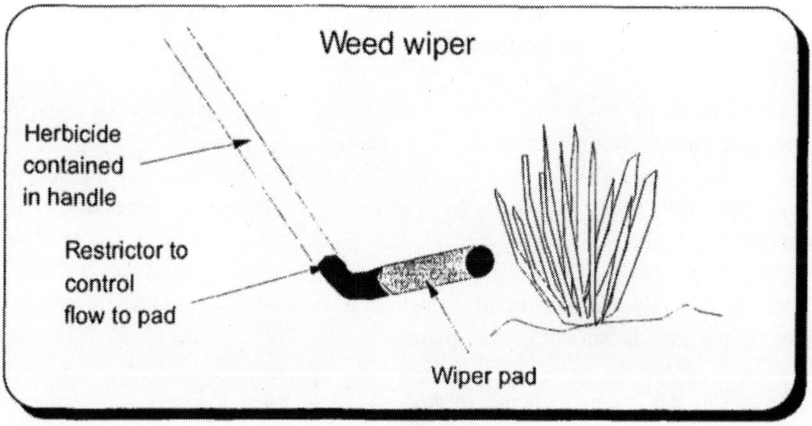

The weed wiper has been used to minimise the possibility of drift. The herbicide mostly used is glyphosate, which is readily absorbed and redistributed within the weeds. A major problem is that the absorbent surface can be blocked by dirt or other debris collected from the weeds. The weed wiper is mainly useful when controlling small patches. of weeds, for example along pathways in plantation crops, or where weeds are taller than a crop, such as groundnuts or strawberries, and where the herbicide must be applied on very localised areas. The chemical is only applied to the target by direct contact and the minimum quantity is used. Visual demarcation of where the herbicide was applied can be achieved by adding a colour dye to the herbicide.

After use the wick must be washed carefully where the contaminated water from the wick will cause no damage. Some users keep the wick, after washing it, immersed in a bucket of water. Old wicks need to be disposed of carefully to avoid the herbicide residues affecting local vegetation.

Dust application

In the past, dusts were used much more on tropical crops than at present, because they were sold ready for use; they did not require water for dilution or mixing. Unfortunately, in some tropical areas some farmers mix emulsifiable concentrates or other spray liquids with locally obtained sand or clay to make their own 'dust'. This is not recommended for several reasons including the possible degradation of the active ingredient due to the properties of the carrier dust. There is also an increased hazard to the operator who unfortunately often mixes and applies the dust with bare hands.

The application of dusts has a number of problems. These are:

a) dry particles need to be extremely small to stick to crop leaves, but such small particles can be inhaled by the user.

b) dry particles are easily blown away by the wind so coverage of crops is poor, and environmental pollution is considered to be high.

c) the low concentration of pesticide in dusts increases the cost of transportation to farms.

d) the pesticide is not easily transferred into the foliage so they are mainly effective against insects crawling over plants.

e) due to the irregular shape of dust particles, often a local source of clay, they are less easy to measure in contrast to the flow of liquid through a spray nozzle.

f) storage and transport problems arise as dusts must be kept completely dry; otherwise the particles will aggregate and form large lumps, that are difficult to disperse.

The main use of dusts now is admixture with grain after harvest to prolong storage and reduce insect damage. Dusts can also be used in stores to treat harbourages of pests. In the field, dusts are mainly used in horticulture, where crops are grown in polythene tunnels. Dusting of these crops instead of high volume sprays avoids increasing the humidity which could increase the risk of disease infection. Dusts have also been used to protect young maize and other cereal crops from early stalkborer infestation, but in this case and other uses of dusts, farmers now prefer to use microgranules. These are very small granules which are too large to be inhaled.

Several types of duster have been designed. These include those with a fan, bellows or bicycle type plunger air pump to create an airstream to project dusts towards a crop. Others are a simple container with a several small holes at one end - a pepperpot - to shake the dust particles. Compressed air dusters, similar in design to a compression sprayer are also available with a plastic or stainless steel container. In many situations an old sock or cotton bag is used as a dust container and this is shaken or hit with a stick to distribute a cloud of dust particles over plants.

Some of the commercially available dusters are designed to sit on the operator 's chest. This means that the lid of the container is just below the operator 's nose, so inhalation of particles and toxic vapour is likely. These dusters should not be used. Preference should be give to equipment that is mounted on the operator 's back, like a knapsack sprayer, with ducted delivery.

Small hand-operated dusters with a rotary fan and ducted air flow are useful in treating potted seedlings in nurseries.

In general the use of dusts is not recommended; where possible use granules or sprays as these formulations are more efficient in control of pests.

Granule application

Some of the most toxic pesticides are formulated as granules as they would be too hazardous to apply as sprays. Most granule products are intended as soil treatments, either to the surface around plants or by incorporation with the surface layers of soil. By formulating as a granule, the release of the pesticide can be extended by coating the granules with various types of polymer. Coating of granules with polymers makes application easier, as the coated granules usually have a more regular shape and flow through the metering system, especially if they are also coated with graphite that act as a lubricant. Other small granules, or microgranules, are intended for foliar treatments, where the granules fall into the foliage, for example on maize or rice, on which the granules are retained by the whorl of leaves.

Differences in the size of granules and their bulk density, ie their weight in relation to the volume of granules, make it more difficult to apply them compared with sprays. The size of granules should be indicated on the package. Table 7.1 indicates the range of particle sizes in relation to different sieve sizes. This variation in particle sizes means that each granule applicator should be carefully calibrated each time it is used with every granule product.

A granule applicator consists of a hopper, a container that should have a sloping floor to facilitate the movement of the granules to the outlet. The hopper should always have a tight water-proof lid to keep the granules dry during application. From the hopper, granules are metered to the outlet. The best metering devices have a rotary action to move the granules from the floor of the hopper and drop them into the dispensing unit. Many granule applicators have a simple adjustable opening as a metering device, but these can be easily blocked by granules forming a bridge over the opening. Some hoppers are fitted with an auger or agitator to keep the granules flowing, but care is needed to avoid breakage of the granules.

Table 3. Characteristics of granules according to sieve sizes.

Sieve size ISO	***Opening µm***	***Number of granules per gram***	***Approx. numbers of granules per cm²***
8/16	2380/1190	320	0.032
16/30	1190/ 595	2530	0.25
30/50	*5951250*	27110	2.7
45/140	354/ 105	167800	16.8
100/200	149/ 74	1367500	137

Rotary metering devices are ideally driven by a wheel so that the amount of granules dispensed is related to the fonvard. speed of the equipment. By choice of the size of the rotor and its speed of rotation, different application rates can be selected. Care is necessary to make sure that the metering device does not grind the granules into dusts.

On most granule applicators the granules fall from the metering device by gravity through a tube. In this case, the delivery tube must be as vertical as possible. If there is any horizontal travel, the movement of granules needs to be assisted by an airflow. The end of the delivery tube may be a circular opening or restricted by a baffle plate or 'fishtail' to deflect the granules into a wider swath. Some equipment granule applicators are attached to planters so that the granules are applied to the side or below seeds.

The simplest granule applicator are like the dusters, a pepperpot type shaker. Some granules are supplied in a box or bag with a tear-off strip to expose the holes for application direct from the container. Where there is a flexible bag dispenser fitted with a metering tube, such as a 'horn seeder, it is possible to swing the tube in a figure of eight and broadcast the granules over a wide swath.

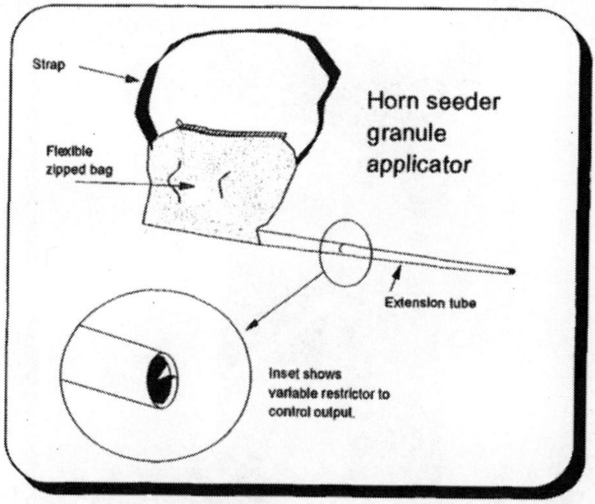

Figure 11 Pepper pot applicator made from plastic bottle to treat maize with granules for stalk borer control.

Spot treatment of granules, for example alongside individual cabbage plants, can be achieved with a tube type applicator with a trigger that operates a valve at the bottom of the container. Each movement of the valve releases a measured quantity of granules to a small area below the opening. Larger quantities are dispensed using a knapsack hopper with a lever-operated cup mechanism to measure individual doses of granules. These have been used to treat bananas to control burrowing nematodes. Chest mounted equipment should never be used.

Fig. 12 Spot applicator.

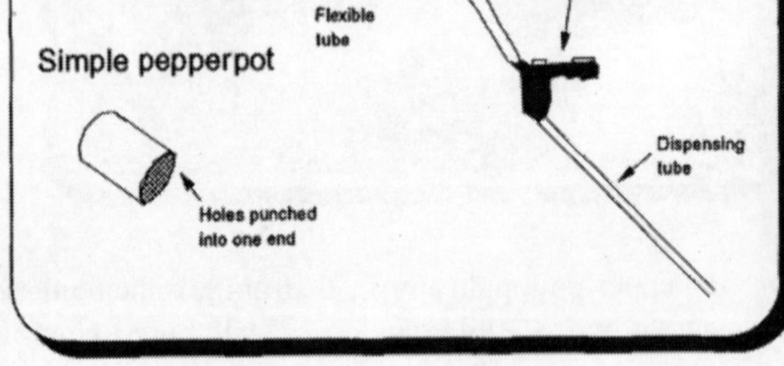

Granules have been used extensively in rice cultivation in the Far East. In most cases a motorised knapsack blower is adapted to apply the granules.

The delivery tube may be extended by a series of short interlocking rigid sections to form a pipe from which the granules are blown through holes along its length. Alternatively there is a long flexible plastic tube inflated by the air pressure from the blower. This tube may be up to 40m in length and is held by two operators, one on either side of the rice field. Instead of holding this tube rigidly, they move it up and down, while walking along the edges of the field. The granules are blown out of a series of small holes, along the length of the flexible tube, each hole having a small flap on the inside edge.

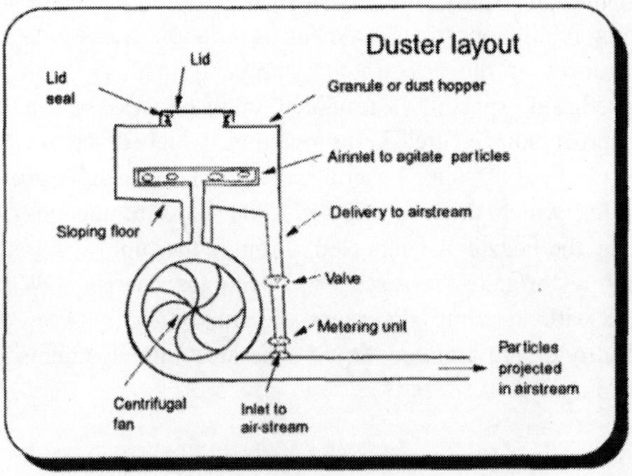

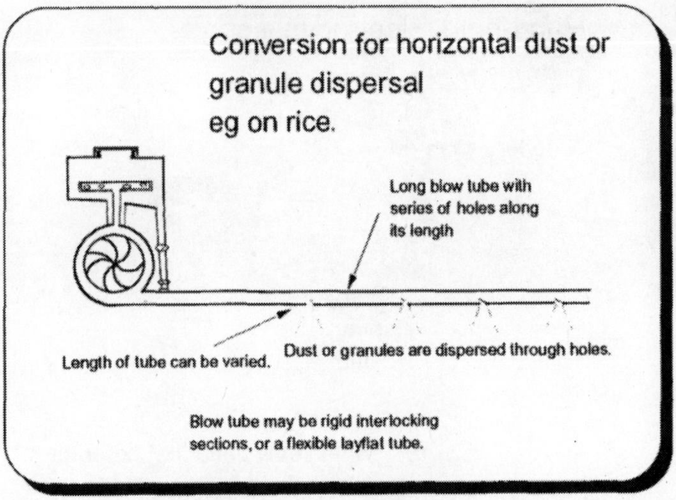

Electrostatic sprayers

Improved deposition of a spray on exposed foliage can be obtained if the droplets are electrostatically charged. The effect of a cloud of charged droplets is to induce an opposite charge on the nearest earthed surfaces that assists in deposition, even on the undersurface of leaves. However, this improved deposition does require sufficient airspace around plants for the charged spray. In practice when plants are close together along a row, and when branches of adjacent rows grow across the inter-row, the charged spray may not penetrate down through a crop canopy, so the improved coverage would then be confined to the upper part of the plants.

Several systems of charging spray droplets have been examined, but only one method has been used commercially on small-scale farms in some areas of Africa and South America. This is known as the 'Electrodyn' sprayer. In this system, an insecticide, such a synthetic pyrethroid, specially formulated in an oil, was sold in a container, the 'Bazzle', with an annular nozzle fitted in the opening, which is screwed into one end of a long tubular handle. Four 'D' size 1.5 volt batteries in the handle provide power to a high voltage generator which then connects 25,000 volts to the nozzle. A counter electrode ring around the nozzle is connected through the tube to a trailing earth wire. Liquid is fed by gravity through the nozzle and becomes charged. When it is emitted around the annulus, with a strong divergent electrical field between the nozzle and earthed counter electrode, the charged liquid separates into ligaments that then break up into the charged droplets.

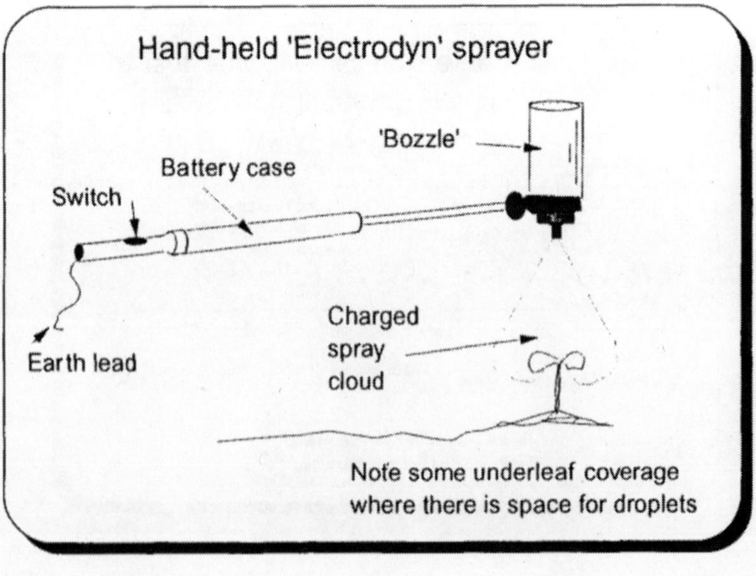

The nozzle has no moving parts and requires extremely little energy to operate, less than 0.1 watt. Four batteries should be effective for at least 60 hours. The sprayer is used in much the same way as a spinning disc applicator, with the nozzle held downwind of the operator, and about 0.4 - 0.5m above the crop. The 'Bozzle' is inserted into the screw ring at the end of the handle; the cap is then removed and attached to the handle while spraying. The earth wire is unwound and allowed to trail while walking through the crop. The 'Bozzle' is then invelied and an electrical switch held open continuously during spraying. At the end of spraying, the stick is turned over, the switch released and the nozzle allowed to touch a plant to discharge any electrostatic charge on the nozzle. The cap can then be put back on the 'Bozzle' and any chemical remaining in it returned to the store.

As the droplets are charged there is less downwind displacement of the spray, so it is usual to operate the equipment with narrow swaths of 1 - 2 metres wide. A particular advantage of the charged spray is that very little pesticide is wasted on the soil so many soil-inhabiting insects, such as beneficial ants, spiders and predatory carabid beetles are less affected by the spray programme. This reduction in wastage of chemical enabled some insecticides to be used at lower dosages than when applied with a knapsack sprayer.

The 'Electrodyn' sprayer was introduced to treat cotton, mainly with pyrethroid insecticides, applied at less than 1 litre per hectare. Due to the cost of the oil-based formulation, it was essential to restrict the system to ultra-low volume applications. Unfortunately relatively few pesticides are sufficiently active in a few grams per hectare and can be formulated in so little liquid. This limit on the number of pesticides available for spraying with the 'Electrodyn' has prevented its wider adoption, particularly where integrated pest management programmes need particular pesticides to be selected in relation to pest monitoring in the crop.

Seed treaters

Seed treatment is an important method of protecting crops during the early stages of seedling development. Some pesticides are phytotoxic to germinating seeds, so careful selection of the active ingredient and its formulation is necessary. Some products cannot be held in sufficient quantity on the seed coat so some seed distributors use pelleting techniques to coat seeds with an inert carrier to hold the pesticides. Usually a strong colour dye or substance which tastes very bitter is added to the pesticide to show that the seed has been treated and to discourage birds and other organisms eating it.

Seed treatment is often done with large equipment centrally by the Company selling the seeds, but there are seed treaters that can be used for small batches of seed, for example the 'Rotastat' applicator operated with a bicycle chain drive.

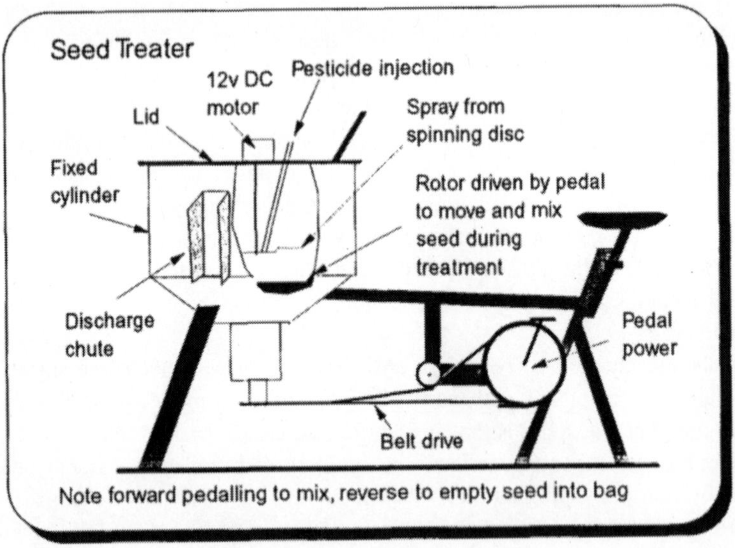

In this equipment, seed is placed in a drum, with a spinning disc mounted in the lid. With the lid closed, one operator sits on the saddle and turns the pedal to move the seed around the drum. The pesticide, formulated as a liquid, is then injected by a syringe on to a spinning disc powered by a 12 volt car battery. When the seed has been well mixed with the pesticide, the pedal is reversed to push the seeds out of the drum and directly into a sack.

Very small quantities of seed have also been treated by placing seed and pesticide in a container with a tight lid, and the shaking or rolling the container until the pesticide has been well distributed on the seeds.

Injection

Certain pests at the top of tall trees have been controlled by injecting a systemic insecticide, such as monocrotophos, at the root zone or low down on the main stem. On oil palms and coconuts, the procedure is to drill a hole between the base of two frond butts, angled 45° downwards into the stem, using a 10 - 15 cm long drill. The insecticide is injected into the hole as soon as possible and the hole covered with a fungicide paste. An alternative method is to put a plastic plug impregnated with pesticide into the hole. With ce1iain formulations of systemic insecticides, painting the stem, without the need to drill a hole, can be an effective application method on some trees.

Chemigation

Many farmers with irrigation systems, including drip irrigation, like to use their method of distributing water to crops to apply pesticides as well. The main problem is that the volume of water applied is extremely large (>5000 litres per hectare), compared with the usual quantities in pesticide application (<500 litres per hectare). The irrigation system needs to have a positive displacement pump and a one-way valve that ensures the water supply cannot be contaminated by a backflow of pesticide. Some systemic insecticides have been effectively applied by chemigation, while most soil applied chemicals are also effective. Where possible the pesticide is fed into the irrigation water for the shortest cycle.

Foggers

Space treatment of buildings in which produce is stored requires very small aerosol droplets that will remain airborne in the store for a long period. As many of the insect pests of stored grain are active in the evening, insecticide treatments are preferably timed in the late afternoon so that the store can be kept closed for at least six hours though the night. Where there is a large warehouse it is usual to start fogging at one of the building and move away from the fog towards a door at the other end of the building. Alternatively if there is access to the building at more than one place, the doorway can be sealed and fog introduced from outside through a small duct at each door. Some buildings have 'fogging ports' at intervals along the side of the building. Ventilation systems inside a building may be effective in distribution of a fog.

Fogging should always be done by trained operators. Two types of aerosol can be used -a thermal fog or a cold aerosol. As aerosol droplets are small enough to be inhaled into the lungs, the trained operator MUST always wear a respirator and other protective clothing, while fogging inside buildings. A high concentration of fog particles in an enclosed area can be easily ignited by an electrical spark, so all electrical switches should be off and naked flames, such as gas pilot lights, extinguished before fogging.

Fogging is sometimes used as a space treatment outside buildings, for example in rubber and cocoa plantations, but if this done, there must be calm conditions with a temperature 'inversion'. Any air movement will rapidly disperse a fog. Never direct a fog into wind as the pesticide will be blown immediately back at the operator.

Thermal Fogger

Most of the small, hand- or shoulder- carried thermal foggers have a petrol driven pulse-jet engine. The unit has two tanks, one for the pesticide and a smaller tank to hold petrol. There is also a manual or electrically operated air pump which is used to pressurize both the tanks, on which the lids must be fitted airtight. To start the engine, the operator opens the fuel tap, pumps air, and also depresses a switch to connect the battery operated ignition system. A spark plug in the combustion chamber provides initial ignition. Once ignition is established, the hot exhaust gases will ignite subsequent charges of fuel and air in the combustion chamber, so the electrical switch is no longer used to provide a spark.

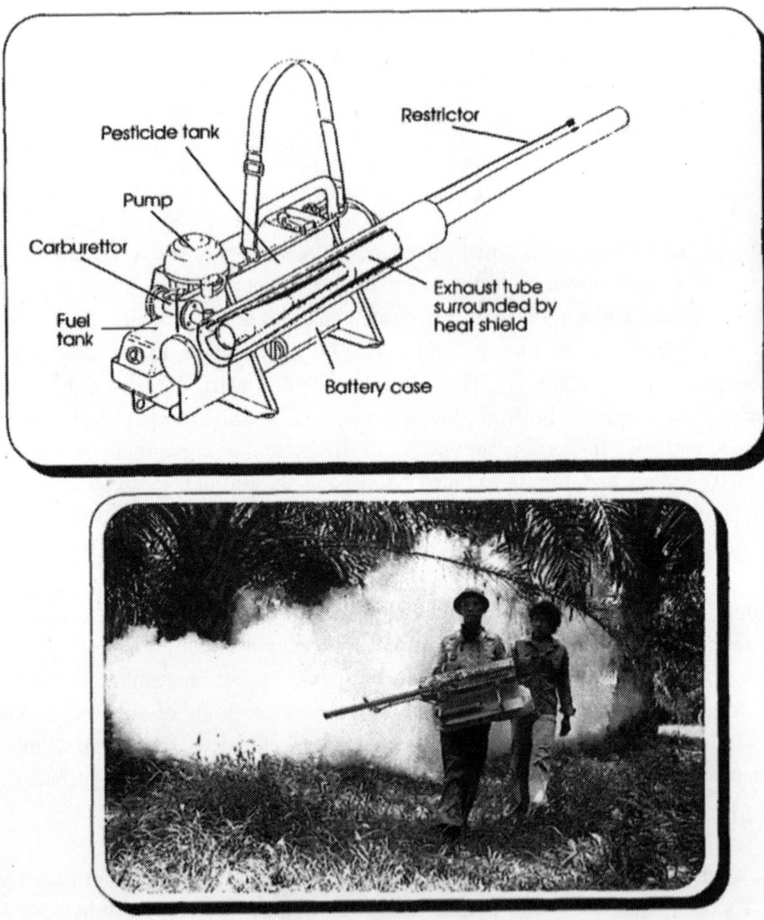

Figure 13 Hand-carried thermal fogger.

The engine should run with a loud noise (over 90 decibels) of about 80 pulsations per second, so the user should wear ear defenders to protect the ears. When the engine has warmed up, the valve on the pesticide tank can be opened to deliver the liquid, usually an oil-based formulation to the restrictor and nozzle positioned near the open end of the long exhaust pipe. The pesticide is vapourized by the very high temperature (>500°C.) in the exhaust, but condenses rapidly when emitted into the air to form a dense fog, with particles generally smaller than 1 Sμm in diameter. If the restrictor selected allows too much liquid to be injected into the hot gas, there will be insufficient vapourization. This will produce some large droplets of pesticide that will fall out on the floor close to the fogger. It is important therefore to select a restrictor that will reduce the risk of this occurring. Some pesticides are degraded by the high temperature, although the chemical is exposed to it for a extremely brief period.

A thermal fog is warmer than the ambient air temperature, so the particles will rise though the air space. These fogs are therefore usually emitted close to the ground level of a store. A fog will normally fill all the airspace around stacks of produce, but it will not penetrate into stacks of grain in contrast to a fumigation.

On completion of fogging, the pesticide valve must be closed and the engine allowed to run until all the pesticide has been emitted from the nozzle. Great care is needed to avoid pesticide getting in the exhaust pipe without the engine operating, as the pesticide liquid may be ignited. In consequence, always ensure that you do not run out of fuel while fogging. When using a thermal fogger, an appropriate fire extinguisher should be available in case of an emergency. If the fog from a thermal fogger ignites a flame will be emitted. If this occurs the pesticide valve should be closed immediately, when the engine exhaust gases should extinguish the flame. Never leave a thermal fogger operating while unattended and do not operate a fogger too close to a wall or other obstruction (at least 60cm clearance is needed between engine and object) which may prevent air flow and cooling. If the fogger overheats it may be permanently damaged. When used in the open never direct fog upwind as a gust of wind could force the vapourised formulation back into the combustion chamber and ignite it.

Most of the hand-carried machines require one litre of petrol per hour of operation. Never attempt to refill either the petrol or pesticide tanks while the engine is hot. Always refill the fogger in the open using a filter funnel.

Thermal fogging should only be done by trained personnel who observe all the safety instructions.

Cold Fogger

An aerosol can be produced by shearing spray droplets in an air vortex without heat, hence the name cold fogger. A range of different cold foggers have been designed with a electrical, petrol- or propane powered engine. The smallest hand-carried equipment has either an electric motor or a two-stroke engine.

The engine drives a fan to push a large volume of air at low pressure through a vortical nozzle. In this nozzle, liquid is injected into an airstream that causes primary break-up into droplets. These droplets then enter an air vortex, created by a set of fixed vanes around the nozzle orifice. The vortex spins the air laden with droplets so that the liquid is sheared into the aerosol. In contrast to a thermal fog in which the pesticide is diluted in an oil or kerosene, a cold fogger is used normally with an ultra-low volume formulation of pesticide.

An electrically operated cold fogger can be installed in a warehouse with a timer so that it can switched automatically when the building is not occupied. It is vital that the user reads and understands the safety precautions supplied with the machine, as inappropriate and careless use may lead to fire, explosion or electrical shock.

No repairs should be made to the electrical side of the machine except by suitably qualified staff The use of extension cords or leads introduces a new hazard to the user and any such leads which are used should be checked by a competent electrical engineer. The extension facility used must be of the correct size and appropriate, not only the voltage and current, but also take into account THE LENGTH OF LEAD BEING USED. This is often overlooked by users, leading to poor machine performance due to voltage drop in the lines and also providing an increased hazardous situation for the operator. These machines often use oil-based formulations which can attack and damage extension leads or wiring, so any of these components should be oil resistant.

Whether a cold or thermal fog, some aerosols suspended in air can be capable of causing an explosion, ifthere is a source of ignition and the concentration of fine particles in the air is greater than a certain level. Some manufacturers state in their safety precautions that an acceptable oflevel liquid in the atmosphere is 2.7 litres per 1000 cubic metres. This provides a safety margin of 5 to 1. Paiiicular care should be taken of any gas or oil pilot lights or sparks that could be caused by electrical controls and all of these should be extinguished and turned off during treatments.

Any damaged or worn electrical leads or controls must be repaired to replaced immediately and no machine should be used with these components in a damaged condition. The machine draws air across the motor for cooling and also draws in air to deliver to the nozzle, so any inlet filters should be kept clean and the suction area of the motor kept clear.

Hand carried versions of this equipment should not be operated unattended.

Summary

There are a number of situations where conventional spraying is not appropriate. Specialised equipment is needed for seed and space treatments. In addition it may be necessary to apply the pesticide dry as a granule or in the case of some translocated herbicides to smear the herbicide on to isolated weeds.

CHAPTERS

MINIMUM SPECIFICATIONS FOR EQUIPMENT

Frequently there is a limited choice of pesticide application equipment available in a local store or in a particular country. The bulk purchase of equipment by Government Tender Boards and others is often based on price; the lower the unit cost, the more machines that can be purchased for a fixed amount of money. Unfortunately the cheapest equipment is seldom very durable and lack of spare paiis may mean that the user is unable to repair the machines. This means that sprayers are often used when they leak or have some other defect which makes application of the pesticide very inaccurate and more hazardous.

Some countries have published specifications on pesticide application equipment. Some of these specifications do provide standards for manufacturers to meet in terms of consistency of their product, but it is only recently that Health authorities and others have sought changes in the design of equipment to improve safety in its use. The operator of manually carried pesticide application equipment is much closer to the pesticides than those using tractor-mounted equipment, and in consequence is much more exposed to pesticide contamination. This is liable to occur not only during mixing the spray, but also during the much longer period of application. The user often walks directly onto areas treated with pesticide. Even though the pesticide is now diluted, the operator can be wet with spray over a long period. As the user is more exposed to pesticide contamination in this way, there is an urgent need for better quality equipment, that does not leak and is safe to use.

Since 1964, the World Health Organisation has published specifications in respect to equipment used to apply pesticides to control the vectors of human diseases. More recently at the instigation of health authorities, some countries have published new specifications designed to provide minimum standards of quality and minimise operator contamination.

Examples of the specifications of compression, lever-operated knapsack sprayers, motorised mistblowers and other equipment described in this Bulletin are given in a series of Appendices.

All those responsible for funding and purchasing of equipment for control of pests should encourage the use of these specifications, so that the pesticides are applied with more safety to the users and the environment.

CHAPTER 9

SAFETY

All pesticides are marketed, because they are TOXIC to certain pests. Unfortunately pesticides are seldom very selective, so they are also toxic to some extent to man and other organisms. Pesticides are classified according to their toxicity, usually evaluated initially against rats as an example of a mammal. In each chemical group of pesticides, there is a wide range of toxicities to different organisms. Some of the earliest organophosphate insecticides, such as parathion are extremely toxic, whereas others, such as malathion, are less toxic. The risk or danger of being poisoned, that is the HAZARD of using a particular pesticide will depend on the extent of EXPOSURE of the user to the pesticide. The hazard can be reduced by the way the pesticide is packaged, sold and subsequent used.

Apart from the risk to the operator, pesticides can adversely affect other organisms so whenever possible the pesticide which is most selective for the pest and least persistent in the environment should be chosen.

Operator Exposure

At present the greatest hazard for the user of pesticides is measuring out the concentrated product to dilute with water, particularly with manually carried equipment, when small quantities are needed. Often the user has to pour a few millilitres of liquid from a container into a small measuring cup. This often results in spoilage over the user's fingers. Effo1ls to measure dry powders with a small cup or spoon seldom give accurate dosages of pesticide and particles can puff up into the user's face. One of the original aims in developing spinning disc sprayers for ULV application was tq provide prepacked containers fitted directly to the sprayer to avoid the risks in mixing pesticides.

Reducing Hazard

Hazards at the mixing stage when using pesticides have been recognised and several changes are occurring:

Agrochemical companies are redesigning packages; for example with liquid formulations, containers have a 45 or 63 mm diameter opening to make it easier to pour the Product without splashing. The aim on the larger types of sprayer is to have a closed system of transferring pesticides, so that with a standard opening of 63 mm on all containers larger than one litre capacity, the container can be directly attached to the sprayer. However, for the smaller quantities used in compression and other knapsack sprayers, there are now packages with a measure moulded into th'e container. When a liquid container is empty, the residue on the inside should be washed out several times and the washings added to the sprayer, before completing the dilution of the final load.

When using dry formulations, some manufacturers are now .marketing sachets made from a water soluble plastic. When the outer package to keep the sachet dry has been removed, the sachet can be added directly without opening it to some water in the sprayer. Some products are aggregated as a tablet, that is added directly to the sprayer. These systems assume the user will always need an exact number of loads of the sprayer, and not have any diluted pesticide to dispose after spraying a crop. Users need to be trained to avoid using too much pesticide so, to avoid a disposal problem, it is better to leave a small part of a crop untreated.

The user must **READ THE LABEL.** The label provides information legally required for REGISTRATION of the Product in any particular country, Unfortunately a large amount of information is needed for registration, and is sometimes given in more than one language. This makes it difficult to read the label and many users are unable to understand all the details. To overcome some of these difficulties, there is increasing use of small diagrams, known as PICTOGRAMS and additional information is often on a separate leaflet.

Pictograms were derived initially as the international road signs. Using the same idea there is a series of diagrams to indicate HAZARDS, such as the toxicity of a Product. Other pictograms give advice on protective clothing, needed at different activities during application, such as when handling a liquid concentrate. Others are warning signs. Users of pesticides need to be trained so that they understand the full meaning of each pictogram. Manufacturers put these pictograms on Product labels, but in some cases they are too small and not easy to see.

In some countries, there is legislation that demands that each pesticide label should be in a particular colour to indicate the toxicity of the Product. In some countries the most highly toxic pesticides have a purple label. As there is less toxicity, the colour is red, orange and then green for the least toxic.

Most countries also publish a list of Products that have been registered. Distributors of pesticides should have the latest edition for their area, so that appropriate advice is available to customers.

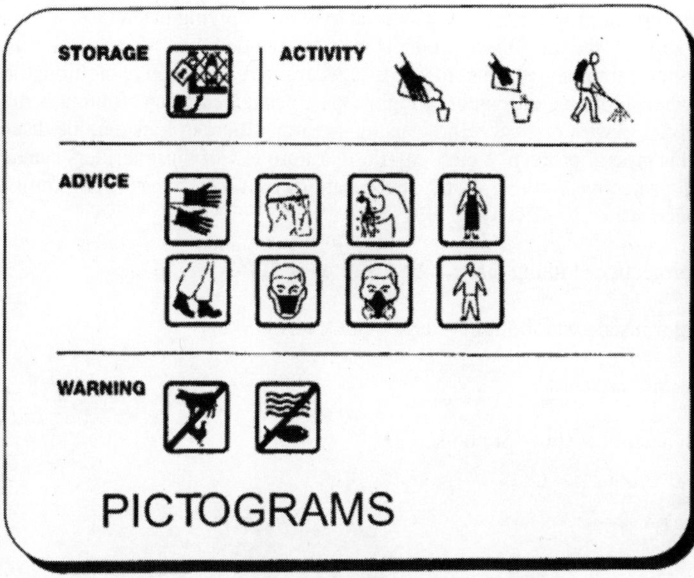

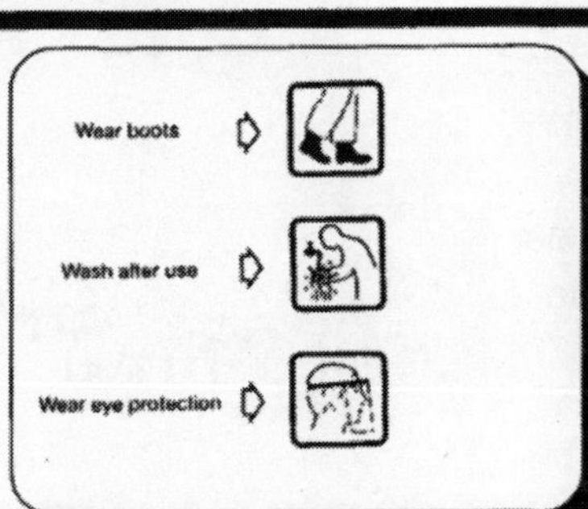

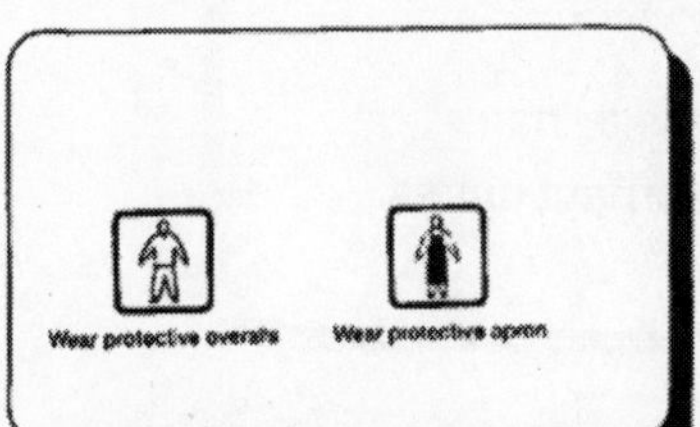

Pictograms are symbols
to aid safety in pesticide use

Preferred size is 15 x 15mm

Pictograms should relate to the safety text on the Product label.

Protective Clothing ‘

In any pest management programme, it is essential to avoid applying pesticides, if possible. When they have to be applied, the most toxic products must be avoided. In a hot tropical climate, no user of manually operated spraying equipment wants to wear the full protective clothing that has been recommended when applying extremely or highly toxic pesticides. Such clothing is impermeable to aqueous sprays, and with manual effort causing sweating, there is considerable discomfort when wearing this type of protective clothing. Furthermore few of these farmers can afford to purchase special protective clothing. Some specifications for protective clothing for use in hot climates have been issued by GIF AP.

The minimum protective clothing that can be expected is as follows:

1) long trousers, with bottom of legs outside any boots.

2) long sleeved shirt

3) non-absorbent shoes or boots

4) a hat

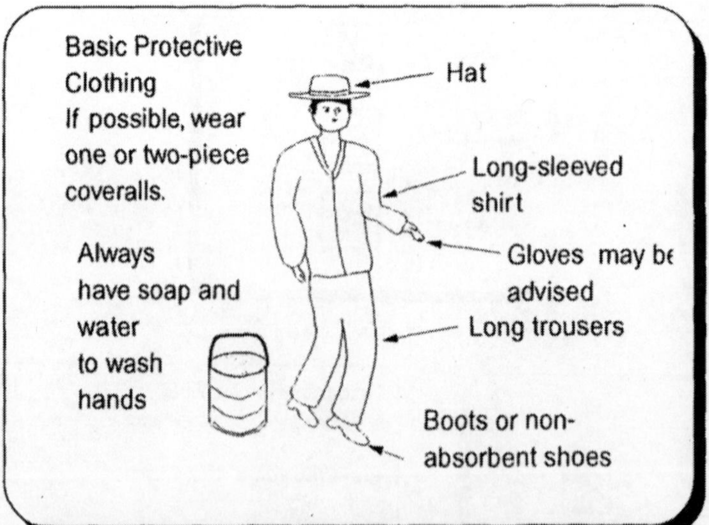

Some national dress gives good coverage of the body and can be used. Always make sure that any holes the clothing are repaired.

Coverall

If the user can purchase some additional clothing, then a light heavy-duty closely woven cotton overall or two-piece coverall should be used. Coveralls should not have pockets, and any zip should be covered by material. The end of the trousers and sleeves may be elasticated to reduce the risk of contamination on the inside. Some have a hood. All coveralls should be washed well regularly with detergent or soap and rinsed several times to ensure no build-up of residues in the fabric.

There are now light-weight plastic disposable overalls, which may be re-used several times. However, they are often considered too hot to wear.

Gloves

In addition suitable gloves must be worn with the more toxic products while opening containers and transferring pesticide to a sprayer. The gloves are usually made of a synthetic nitrile or neoprene rubber or plastic, such as polyvinylchloride (PVC) that is impermeable to water. Gloves made of natural rubber are not recommended as these have poor resistance to many solvents in liquid formulations. Wearing these gloves for any length of time is liable to cause the hands to sweat, so the skin becomes more open to pesticide absorption. Any glove should be checked regularly to make sure there are no holes through which the user will get contaminated. When being washed and rinsed, gloves filled with air can be held under water and squeezed to see if any air bubbles escape through holes.

Solvents, such as xylene, used in some pesticide formulations can penetrate through a glove in minutes, so users must have a supply of clean water readily available so that if a glove is contaminated, the outer surface can be washed quickly. The gloves should always be washed before careful removal to make sure that any pesticide on the outside of a glove is not transferred to the hand.

The glove should be selected so that it is neither too thin, likely to tear and give inadequate protection, nor too thick, when the user has difficulty in manipulating the fingers and has difficulty in opening containers. A thickness of 0.4 mm has been suggested as a minimum thickness. The glove should be long enough (at least 30 cm) to tuck inside the cuff of the sleeves of the shirt to stop any contamination on the sleeve running down inside the glove.

Some of these gloves have a cotton liner to absorb perspiration, and make it easier to remove the gloves. As this liner can absorb pesticide, it must be washed very thoroughly after use. Generally gloves with an absorbent liner are not really suitable for pesticide application and NEVER use a glove made only of absorbent material, such as cotton or wool. Gloves very contaminated with organophosphate or carbamate insecticides should be soaked in washing soda for at least eight hours, rinsed and then dried. Discarded gloves must be destroyed and not left for someone else to use.

Whenever the user does not use gloves, always have a good supply of soap and water immediately available to wash off any contamination of the hands.

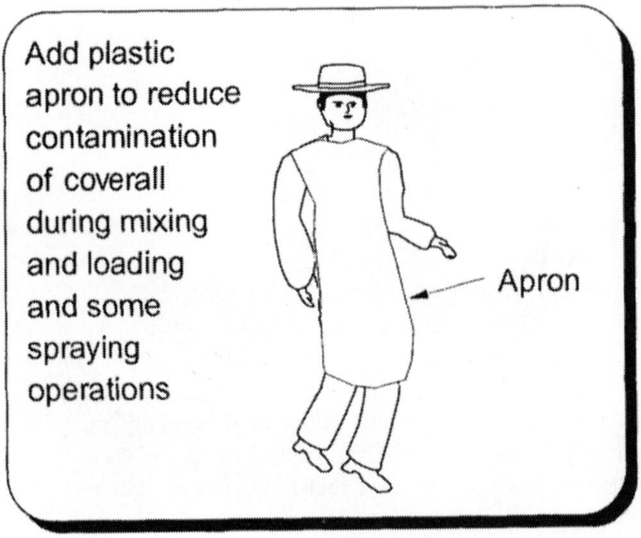

Apron

Where possible the person mixing pesticide sprays should also wear a plastic apron to reduce the risk of the basic clothing being contaminated with concentrated pesticide. Such an apron has also been recommended when the spray operator is walking directly into the area being sprayed to reduce the contamination of clothing that touches sprayed vegetation.

Rubber boots

These should be used in preference to other footwear. The trouser leg should always be on the outside of the upper part of the boots so that any pesticide spilt is not collected inside the boot. Any footwear should be washed before removal to avoid hands becoming contaminated by any pesticide deposit on the outer surface of the boots.

Face Protection

The person opening pesticide containers may splash concentrate even though it is easier to pour from new designs of container. The simplest and coolest way to protect the face and especially the eyes is to wear a face shield. They are made of a clear plastic material and are sometimes easily attached to a plastic hard hat. Alternatively the user can wear safety glasses or goggles to protect the eyes. Some makes of goggles can mist up and reduce visibility.

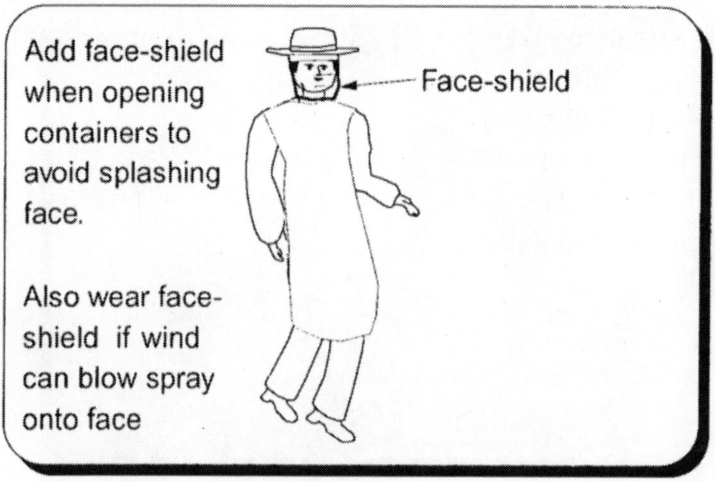

These extra items are expensive, so the alternative is to wear a cloth scarf over the lower part of the face, at least while opening containers. Disposable masks that reduce the area of facial skin exposed to spray deposits are available commercially. They can be used also as dust masks, but they should never be considered as respirators, as toxic vapour can still enter the lungs, especially as most users fail to get a tight fit of the mask around the nose.

When applying aerosols, fogs or very fine sprays inside buildings, the operator must wear a respirator. There are several types, but the most commonly used respirators have a canister or cartridge containing substances which will absorb the toxic vapours of pesticides. It is essential to use the correct cartridge for the type of pesticide being used, and that it has not exceeded its service life. Their use is also very uncomfortable in hot climates due to perspiration and moisture condensing on the inside. They must fit the operator's face correctly, so people with beards will find it is difficult to get an air-tight seal.

When indoor treatments are needed, staff need to be specially trained in. the techniques, especially when applying aerosols or fogs. There must be an accurate record of when respirators with cartridges are used and the cartridges must be destroyed after the limited period of effective service life. If in doubt, always use a new cartridge. After use, the cartridge should be removed and the rest of the respirator face-piece washed and dried, before being stored in a dark place for future use. Rubber face pieces will be less prone to become sticky if dusted with talcum powder.

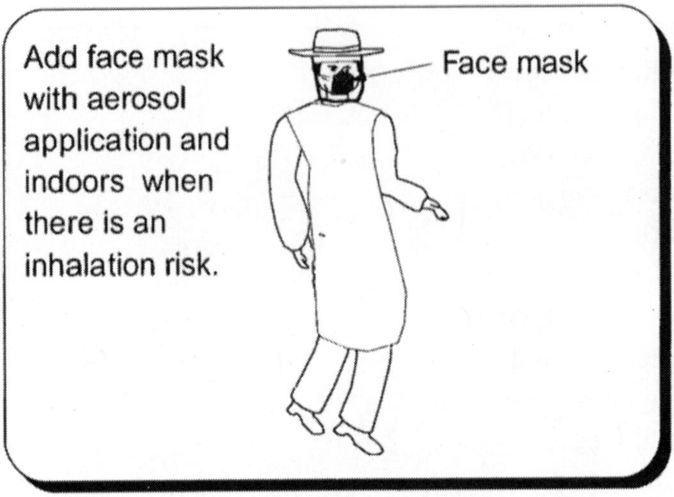

Hat

Hair is a very effective collector of small particles, so it is advisable to keep the pesticide user's head covered by a hat at all times. A wide-brimmed hat also acts as a sunshade to keep' the person cooler. In one area a light hat with a peak is available. It has a large fringe at the back that covers the neck and sides of the head and can be joined under the chin to protect the front of the neck. Covering long hair is also important where there is a risk of it being caught or sucked into machinely.

Farmers should try to avoid applying pesticides in the hottest part of the day, but sometimes when it is cooler, there is no wind to carry the spray away from the operator. Even if an early start has been made, the farmer may wish to continue until the whole area requiring treatment has been completed.

Ideally the material used in the hat should be as non-absorbent as possible. When spraying up into trees, a waterproof hat is essential.

Figure 14 Example of Protective clothing.

Before spraying in the field

There should always be a complete check on the sprayer before commencing to spray with pesticides. It is particularly important to ensure there are no leaks from any part of the sprayer and that the sprayer has been calibrated with water so that any adjustments can be made before the equipment is contaminated with chemicals. In the case of some formulations a sample of the actual spray has to be applied, if the flow rate is significantly different from water or appropriate inert liquid.

Always check that there will be clean water for washing easily available while spraying. Also always provide separately sufficient drinking water to replace fluid loss from the body due to sweating. Make sure the drinking water is protected and cannot get contaminated by pesticide.

Only remove from the store, sufficient pesticide for the day's programme. Never transfer pesticide to another container.

Lastly the person applying pesticides should warn his neighbours or other bystanders where pesticides are to be applied, so that they can keep upwind and avoid being contaminated. Make sure children will not enter the fields while applying pesticides. A warning is also needed if there are bees being kept in the area.

During spraying

The person applying pesticides should always observe the direction of the wind and try to have the nozzle downwind of the body. Unfortunately most users of manually carried sprayers tend to hold the nozzle directly in front of their body and walk into the spray or the treated foliage. Some even swing the nozzle from side to side to increase the swath. These practices also result in heavy contamination of the legs, feet and other frontal areas of the body.

It is much safer to treat the crop plants or weeds on the downwind side of the operator. Adjustment of the angle of the nozzle on the end of a lance can be done easily to provide the right swath of fan nozzles. Under field conditions the wind will always carry small droplets away from the nozzle, so that if the operator is upwind there will be less contamination, and any risk of inhaling droplets is minimal. The start of a treatment should be on the downwind side of a field and progressive passages across a field should be upwind of the previous swath.

Some equipment has been designed with a side-boom or an extended lance with several nozzles held down-wind to increase the swath, when treating crops like rice, yet reduce operator contamination. On other row crops, a boom has been mounted behind the sprayer so that the operator walks away from the spray.

If a nozzle is blocked during application, check whether the filter is blocked an clean it if necessary with water. Replace a blocked nozzle with a new nozzle and clean the blocked nozzle later, to avoid delaying a treatment.

Manual effort needed to apply pesticides is tiring, so the operators should take rest periods. This will reduce the risk of mistakes occurring.

After Application

If possible put a notice at the side of the field to warn others not to enter the field. With some pesticides there is a minimum period before entry, even to sample the pest population. Similarly there is a minimum period between the last application and harvest. Any unused pesticide must be returned safely to the store which should be locked to prevent children and others entering it. Many small-scale farmers will not have a separate store, so they should keep the smallest quantity of pesticide as possible on the farm and wherever possible have a lockable cupboard in which to keep the pesticides.

Any rinsed empty containers can either be disposed of safely or put back into the store for disposal later. Remember that it is considered that it is impossible to clean a container sufficiently to use it for another purpose. Nevertheless old containers are valued on tropical farms. The best way to clean them after the initial rinsing, is to fill the container with water and, with the lid firmly in place, leave the water soaking in the container for as long as possible. The water can be used as diluent on the next spray application. The container should be washed in this way at least three times, so that any residue impregnated in the container liner is leached into the water.

When oil-based ULV formulations are used, water cannot be used for cleaning, and where possible the containers should be returned to the pesticide distributor for recycling.

The sprayer should be thoroughly cleaned before being stored. Never leave any residue of pesticide in the sprayer, so rather than just rinse the sprayer tank, it is important to put water in the tank and then pump it through the nozzles to remove all pesticide from the pump, hosing and nozzles. Any residue left inside the sprayer will become concentrated as the water evaporates. This concentrated residue can affect parts of the sprayer such a diaphragms in pumps and check valves, and other seals. Thorough cleaning of a sprayer is also important if the equipment has to be mended, to avoid the person taking the equipment apart becoming contaminated.

Few small-scale farmers have a separate store to keep their pesticide application equipment, so after washing it to remove pesticide deposits, it is important to keep it in a safe place free from the risk of damage. One problem is that rodents are known to chew plastic tubing and damage spray equipment.

The next stage is to remove and clean protective clothing used during the day's application work. Lastly the person who has been doing the spraying should wash well and put on clean clothing.

In a pest management programme it is advisable to keep a record of the names of pesticides used, the quantities applied and the fields and area treated. In some countries it is a legal obligation to keep these records. In addition it is useful to record more precise details of the application technique, such as the volume of spray applied, nozzles used, pressure, weather conditions and any other details that may be useful in assessing the season's results and help plan the next season.

SAFE USE - a Summary

1. Use alternatives to pesticides; when ACTION THRESHOLD is reached choose least hazardous pesticide.

2. READ and UNDERSTAND the LABEL of the pesticide.

3. Wear suitable PROTECTIVE CLOTHING.

4. Check SPRAYER is working correctly; calibrate with water.

5. Follow label instructions for correct dosage and dilution of pesticide; check weather conditions; avoid preparing too much spray to eliminate need for disposal of any surplus.

6. Have a supply of clean water and soap for DECONTAMINATION if required.

7. WASH befol·eeating, drinking, smoking, going to toilet.

8. After application wash equipment and dispose of any waste pesticide safely.

9. Never decant pesticides into beverage, soft drink or other containers.

10. Wash well and wash protective clothing; store clothing, chemicals and equipment safely.

CHAPTER10

MAINTENANCE OF EQUIPMENT

The purchase of equipment to apply pesticides is expensive for the small-scale farmer. The equipment therefore needs to be durable so that the capital cost can be spread over as long a period as possible. However, use of the equipment inevitably causes wear of some parts so routine maintenance, together with careful storage when not in use should prolong its life. All equipment should be provided wit,h an instructional manual which sets out the routine maintenance required to keep it in efficient working condition. Unfortunately some manufacturers provide a separate manual for any engine on motorised equipment instead of integrating the information into one 'manual. The manuals for all equipment should be kept in a safe place, and accessible at any time when equipment needs to be repaired.

All equipment should be cleaned daily after use and stoi-ed in a dry secure place, well away from any food and inaccessible to children and farm animals. The equipment should be protected from rats that can chew hoses. When it is used next, a simple check of a sprayer for any leakage should be a standard procedure before any pesticide spray is put into it. Unfortunately, often under field conditions, users will continue to use equipment until it fails before doing any maintenance.

Washing the Sprayer

At the end of each day's pesticide application, all chemical must be removed from the equipment. Any residues left inside equipment are liable to cause corrosion.

A sprayer tank should be emptied as much as possible, then a small quantity of water added tq_ the tank, which is shaken to wash the inside surfaces. This water is then pumped through to the nozzles and disposed safely. Some countries have regulations regarding the safe disposal of contaminated water. On small-scale farms where the quantities of contaminated water should only be 2 - 3 litres, one way of disposal is on the edge of the field that has just been sprayed. The user of pesticides must ensure that the pesticide cannot contaminate local water supplies.

The nozzles can be removed and each part - the tip, filter and cap washed separately. This is particularly important as small residues of particulate pesticide can dry and alter the size and shape of the orifice. The small orifice of the nozzle must not be damaged so if there is any residue of pesticide, it is better to soak it for a short period. A nozzle should neve.r be put to the user's mouth to blow through it. Also never use a hard object such as a nail or stick to remove a blockage. After soaking a nozzle, any residue that is not dislodged by washing, can be removed by using a bicycle pump to blow air through it. An old toothbrush with soft bristles is useful for cleaning nozzles; the brush has to stored carefully after use. If a nozzle has been blocked in the field, it should be checked after washing to check that the spray pattern has not been affected.

Before replacing the nozzle, put fresh water into the tank and repeat the washing of the tank and pumping the water through the hose two or more times. The repeated washing with a small quantity of water is better than filling the tank once. The outside of the apparatus should be cleaned as chemical may be on the surface due to spillage or spray drift.

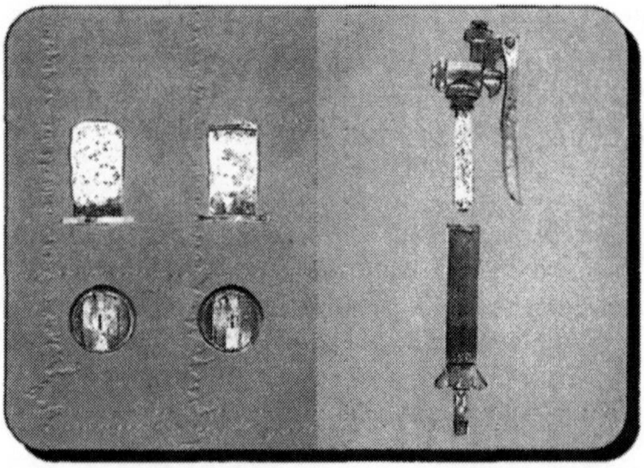

Figure 15 Filters and nozzle requiring cleaning

Other Filters

In addition to the filter in the nozzle body, most sprayers have two or more other filters that need to be checked and washed. The filter in the tank opening is easily removed and washed. There should also be a filter in the handle of the trigger valve, which should be unscrewed to remove the filter for washing. There may be another filter at the inlet to the pump. This is not usually removed, but cleaned by the washing of the tank. If the filter in the tank opening has been kept clean, the risk of blockage elsewhere in the spray line is reduced.

Leaks

Sprayers consist of many parts which are joined together. The joins are the most likely places where liquid can escape. The equipment is designed to minimise the risk of leaks, but this often means there is an 'O' ring seal or washer at the join. When a leak occurs it is usually necessary to replace the seal. When parts have to be taken apart for cleaning, such as the handle of the trigger valve, care is needed to avoid damaging the seal so that it can be replaced when the unit is re-assembled. Any screw connections should be kept clean as small debris can cause a leak. The seals are usually made from a synthetic rubber which can be affected by some of the pesticide formulations, if the equipment is not washed well after use. When a replacement pali is not available, a temporary repair may be possible with a small washer cut from an old cycle inner tube.

The lids of the sprayer tank should always be checked to provide a liquid-tight seal to avoid pesticide spray being spilt down the operator's back.

The hose should be checked to ensure the outer surface has not been damaged. If there is any damage which might result in spray escaping under pressure, the hose should be replaced. Many manufacturers have fitted clips which are easy to fix in the factory, but these cannot re-used. In consequence specifications state that any hoseclip should be replaceable.

Faults with lever-operated knapsack sprayers

Lack of spray is usually due to a blockage at the nozzle or because one of the filters is clogged. If there is still no spray, or the pressure is low, there may be a fault in the pump. On piston pump sprayers, the pump seal may have shrunk and no longer fit tightly on the pump cylinder. As these are now made of a synthetic elastomer, a new seal has to be fitted. On some sprayers the inlet and outlet valves may have swollen due to interaction with the pesticide, especially if the equipment has not been washed well after previous use, so the valves no longer move freely to allow passage of the spray liquid.

Low pressure at the nozzle may also be due to a fault in the air-pressure chamber. On some sprayers, this part is screwed into position, so if incorrectly installed, air or in some cases liquid is escaping. When the chamber is inside the tank, this leak may not be easily detected. Some sprayers are now fitted with a maximum pressure safety valve. If this is set incorrectly, pressure may be lost into the spray tank. If the valve is inside the tank, remove the spray liquid and wash out the tank, before attempting repairs; this emphasises the need to check that a sprayers working correctly by using water first, before starting to spray.

On diaphragm pump sprayers, examine the diaphragm regularly and replace if there are signs of wear. On some machines, if the diaphragm breaks, all the liquid will pour down the operator's legs.

Faults with compression sprayers

If there is no spray, and the nozzles and filters are not blocked, check that the tank has been filled to the correct level, and that the pump is working correctly. A few s.trokes of the pump, outside the tank, should indicate whether it is working properly as on each pressure stroke, the valve should 'grunt', or make the sound of air escaping. When replacing the pump into the container, make sure there is an air-tight seal. Other faults are similar to those described above.

Faults with motodsed knapsack mistblowers

Users of these sprayers frequently ·complain that the engine is difficult to start. This is often due to leaving fuel in the carburettor and engine during storage. As a result, the volatile petrol can vaporise leaving an oily deposit on the spark plug. Generally always stop a two-stroke engine at the end of a spray session by turning off the fuel tap and allow the engine to continue to run until starved of fuel. Always ensure the correct fuel is used in an engine; this information is usually embossed on the fuel tank, for example 1:24 means one part of oil is added to 24 palis of petrol. Storing fuel in the sprayer fuel tank for prolonged periods should also be avoided.

The plastic fuel delivery tube gets hardened and should be replaced regularly. If it is not replaced, fuel could spill on to a hot engine and cause a fire. Similarly the spark plug should be replaced after every 250 hours use, and the plug checked and cleaned more frequently to make sure there is the correct gap between the electrodes. This may require taking the sprayer to the local supplier, who should have the equipment needed to check the engine. A tachometer is useful to check that the speed of the engine is correct.

Other possible faults that prevent the engine starting include a blockage in the air intake and fuel line or the main jet on the carburettor, or the exhaust could be blocked. In the carburettor, the float valve may also stick or condensation of water vapour may require the unit to be dismantled and cleaned. Users should also avoid flooding the engine with fuel.

When the engine has been started, it needs to operate at full speed. If the engine lacks power, check that the choke is closed and that there is no blockage in the fuel line or air intake. The join of the carburettor to the engine should also be checked to ensure there is no leakage of gases, indicated by a whistling noise when the engine is running. If the engine backfires, the spark plug may need cleaning and the ignition checked, but the latter requires trained personnel, if the magneto needs resetting to the manufacturer's specification.

If there is no spray, the most likely cause is a loose fitting lid to the spray tank: Without air pressure the spray liquid will not reach the nozzle if it is held above the level of liquid in the tank.

Faults with spinning disc sprayers

If there is no spray there may be a blocked restrictor, which should be cleaned carefully to avoid damage to the orifice. If the speed of the disc fails to rotate, check that batteries have been placed in the sprayer and in the correct position, with the polarity of the batteries - '+' and '-' as indicated on the battery case. Also check the connections have not become loose or corroded. If the disc speed appears to be slow, it should be checked with a tachometer and also the batteries checked to ensure they are providing the correct voltage.

Faults with hand-carried thermal foggers

One of the greatest difficulties with these thermal foggers is getting the equipment to start. First check the operating manual to ensure the correct starting procedures are followed. Then check that there is fresh petrol in the fuel tank; the tank lids are air-tight and that all the valves are in good condition. Remove the spark plug and check it is clean. The spark plug electrode gap should be adjusted to 2-2.Smm., and checked by operating the start control with the plug grounded onto the machine, being sure there is no petrol or petrol vapour close to the machine during the test. The ignition coil should produce a distinct buzzing sound. If there is no spark, new batteries may be required. Check batteries are installed correctly. Avoid flooding the combustion chamber with fuel so when trying to start after all the checks have been made, open the fuel valve slowly. If the engine has been flooded, close the fuel valve and operate the air pump to evaporate the fuel. After fogging operations, a clean liquid, usually plain oil should be flushed through the equipment before returning it to a storeroom. This will reduce the development of corrosion of metallic components after any water-based formulation has been used.

Deposits that may build up in the exhaust tube can be removed using a wire brush. It may be necessary to dismantle the tube and soak it to soften hardened deposits.

Faults with hand-carried cold foggers

One problem with these sprayers is the metering valve controlling the flow of pesticide to the nozzle. When the equipment is used a 'flushing fluid' should be sprayed through to avoid any pesticide drying inside the valve, thus affecting the output, when the sprayer is used next. The valve should be calibrated regularly to check that the flow rate is correct for a particular formulation, as changes in viscosity will affect flow though the valve.

On electrically driven aerosol equipment, always get a qualified specialist to do any repairs required on the motor. Where a two-stroke engine is used, see information relating to motorised mistblowers.

The pesticide tanks can be corroded or contaminated with slime if left for extended periods in hot and humid conditions. Routine cleaning with a flushing liquid and keeping the tank dry, when not in use is advised.
The inlet filter on the dip tube may become blocked or displaced, so it should be frequently checked and cleaned. If the dip tube is ever replaced, it is important to note that some tubes are cut at an angle of 45° where they contact the base of the tank.

After maintenance the droplet spectrum can be checked very approximately by operating the fogger for 5 minutes whilst standing on a base about 60 cm above the ground, no liquid fall-out should be observed immediately in front of the nozzle and the floor area in front of the machine should remain dry.

Repairs in the Field

Ideally users of pesticide application equipment should have a small toolkit, which can be carried in a box to the field with a few spare nozzles, seals and other small parts. Repairing equipment in the field should be avoided, but sometimes it may be necessary to tighten a part or replace a nozzle so that an application programme can be completed while weather conditions are favourable. Most users will find that they should have at least one screwdriver, an adjustable wrench, a knife and a length of string. When motorised equipment is used, a spare spark plug and a plug spanner are essential. Spinning disc sprayers may require a small screwdriver and a cross point screwdriver.

Summary

Equipment needs to be carefully maintained so that it is operational as soon as it is needed. Manufacturers should provide detailed advice in a manual with each sprayer. Where a 2-stroke engine is used, the instructions should be incorporated into the sprayer manual. All users of application equipment should have a set of appropriate tools.

GLOSSARY

absorption: movement of a toxicant, light, liquid, etc. from a surface into a body; eg. the movement of a pesticide into plant cells.

acaricide: a substance causing the death of mites or ticks.

acetykholine (ACh): a chemical conductor of nerve impulses, formed at the end of nerves, to' conduct the impulse over the microscopic gap between nerves or between nerves and muscles or glands. As soon as the response has been completed, ACh is destroyed by an enzyme called cholinesterase.

active ingredient: the chemical component (actual toxicant) of a formulation that is responsible for the pesticidal effect.

actual dosage: amount of active ingredient (not of formulation) applied to a given area.

acute toxicity: toxicity of a brief, usually single exposure to an active ingredient, usually expressed as LD_{50}.

... dermal: acute toxicity after a single contact with the skin.

... inhalation: acute toxicity from a single contact through breathing .

... oral: acute toxicity following a single intake through the mouth .

adhesive: an adjuvant that helps in making a spray material stick to the sprayed surface, e.g. sticking agent, wetting agent.

adjuvant: material added to improve some chemical and physical property, hence its effectiveness. Includes emuslfiers, dispersing agents, wetting agents, spreaders, foam suppressant, special oils etc.

adsorption: the process by which chemicals are attracted to and held on the surface of a plant or soil particles.

adulterated: any pesticide, the strength or purity of which falls below the standard or quality stated on the label. Also, a food, feed or product that contains illegal pesticide residues.

aerosol: spray with a volume median diameter (VMD) smaller than SOμm; because a large fraction of the spray may contain droplets smaller .than 20μm aerosols can present an inhalation hazard.

agitation: stirring or mixing a substance, e.g. pesticide, to prevent it from separating or settling in the spray tank.

agro-ecosystem: an agricultural area sufficiently large to permit long-term interactions between all living organisms and their non-living environment.

antidote: a treatment given to counteract the effects of a poison.

aphicide: substance used to control aphids (plant lice).

aquatic: (plant, animal) growing, living in or near the water.

arboricide: , substance killing trees and shrubs.

atomize: to produce spray droplets by passing liquid under pressure through a suitable nozzle, or by spinning it from a rotating surface, or by passing a suitable formulation through an electrical force field.

atropine (atropine sulphate): an antidote used to treat organophosphate and carbamate poisoning.

attractant: a substance that lures pests to a location Usually classed as food-, oviposition- or sex-attractants.

avicide: lethal agent use.cl to kill or repel birds.

bacterium (pl. bacteria): microscopic unicellular organism (prokaryotic with ribosomes and both DNA and RNA, surrounded by a strong capsule wall).

bait: food or other material used to attract a pest to a trap or toxicant.

band application: deposition of a pesticide in a narrow band, usually over or alongside a row of crop plants.

basal application: application of a pesticide to the lower part of stems.

biological control: the use of living organisms that are the natural enemies of pests to control the incidence of pests or pathogens. However, some also include under this term the use of any method that depends solely on biological interactions to control pests or diseases, e.g. host resistance.

biological pesticide: a pesticide, the active ingredient of which consists of a living organism, eg fungus, bacterium or virus.

biorational pesticides: biological pesticides include pest control agents and chemical analogues of naturally occurring biochemicals (pheromones, insect growth regulators, etc.)

boom: pipe or tubing with several nozzles to apply chemicals over a wide area at one time.

botanical pesticide: a pesticide obtained from plants, e.g. pyrethrum.

broadcast application: application of chemical pesticide or fertilizer over an entire field rather than only along rows and on individual plants.

broadleaf plants which possess two seed-lobes *dicotyledons)* which serve as a food reserve. In many species, upon germination, hese lobes become green primary leaves, which fall off when true leaves develop.

broad-spectrum pesticide: non-selective pesticide effective against a wide range of pests.

calibrate: to determine accurately the amount of pesticide being applied over a given area by means of a sprayer, or other application device.

callus: parenchymatous tissue consisting of a mass of thick-walled cells that form in response to wounding or infection by pathogens, or by injection of a pesticide into a plant.

carrier: (1) a material serving as diluent and vehicle for the active ingredient

(2) an organism harbouring a parasite without itself showing the disease.

cartridge: the cylinder with absorbing material inserted into a respirator to absorb toxic gases, fumes or vapours.

CDA: controlled droplet application; a spray containing a relatively narrow droplet spectrum.

certified applicator: person qualified to apply or supervise application of restricted-use pesticides.

check plot See control. Sometimes preferred as there may be no control of a pest in an untreated 'control' plot.

chemical name: scientific name for the active ingredient in a formulation.

cholinesterase {ChE): enzyme necessary for proper nerve function; it is inhibited or damaged by organo-phosphorous or carbamate insecticides.

chronic symptoms: symptoms that appear over a long period of time.

cold fogger: a means of producing an aerosol without heat; not a true fogger

common name: (I) a name assigned to a pesticide active ingredient by the International Standards Organization (ISO) or adopted by national standards authorities, to be used as a generic or non-proprietary name for that particular active ingredient only

(2) well-known or popular name for weeds, insects and diseases.

compatible: when two materials can be mixed together with neither affecting the action of the other.

concentration: content of a pesticide in a liquid or dust, e.g. gram/litre.

contact herbicide: herbicide that causes localized injury to plants where it touches them.

control: (1) untreated subjects, or subjects given a standard treatment, for comparison with those given experimental treatment, synonym check

(2) to prevent, retard or eliminate a weed, disease or pest attack, or to reduce their adverse effects to an economically acceptable level.

cover, coverage: the proportion of the surface area of plants, or soil or other treated surface on which an applied material is retained following dusting, spraying, etc.

crop loss the extent to which yield of a crop is reduced by the damage caused by pest infestation.

cultivar: a variety of a cultivated, domesticated crop species. Usually selected by plant breeders to provide growers with special characteristics, such as high yield, good quality, pest resistance, etc.

cultivation: an artificial ecosystem in which farmers' crops are protected as far as possible from natural competition.

cultural control: method of pest control by means of a skilful combination of agronomic practices such as tillage, planting, irrigation, sanitation and crop rotation so that the environment is less favourable for the proliferation of certain pests.

cumulative pesticides: persistent chemicals that can accumulate or build up in the tissues of animals or in the environment (soil, water).

curative pesticide: a pesticide that can inhibit or eradicate a disease-causing organism after it has become established in the plant or animal.

cuticle: a membranous layer on the outer wall of epidermal cells consisting primarily of waxy substances.

damping off: collapse and death of seedling plants resulting from the development of a stem lesion at soil level.

defoliant: a chemical that initiates abscission.

deposit: amount of pattern of spray or dust deposited per unit area of plant surface.

dermal toxicity: toxicity of a material as tested on the skin; the property of a pesticide to poison man or animals when absorbed through the skin.

desiccant: chemical that induces rapid desiccation of a leaf or plant part.

detergent: liquid normally used as a cleansing substance. Some such products may also be used as wetting agents because of their ability to reduce the surface tension of water.

diluent: an inert material added to an active ingredient or technical material to reduce its concentration.

direct drilling: sowing a crop with minimal or no cultivation of the soil since the previous crop; involves application of herbicides to control weeds

dispersing agent: a material or substance that reduces the attraction between like particles.

dormant spray: chemical applied in winter or very early spring before treated plants have started active growth.

dose: (1) the quantity of a pesticide applied per individual

(2) the quantity of a pathogen used to inoculate living hosts, particularly in quantitative response studies.

dosage: the quantity of a pesticide applied per unit area.

drift: movements by wind of fine spray droplets or dry particles (dust) to an area not intended to be treated. exo-drift implies leaving the treated field, whereas endo-drift remains within the treated field.

droplet spectrum: the distribution (by number or volume of droplets) of spray into intervals of droplet size.

dust: finely divided particles (of plant protectant) with or without diluent, for application in the dry condition.

see median effective concentration.

ecology: a branch of biology concerned with organisms and their relation to the environment.

economic injury level: the lowest population density of a pest that will cause economic damage (loss).

economic threshold: the density of a pest at which control measures should be initiated to prevent an increasing pest population from reaching the economic injury level. Note where the economic assessment is difficult treatment may be taken at an action threshold.

see median effective dose.

emetic: something which causes vomiting and is used as a first aid for some types of poisoning.

emulsifier: surface active substance used to stabilize suspensions of one liquid in another; for example oil in water.

emulsion: suspension of minuscule globules of one liquid in another; eg oil in water emulsion.

encapsulated formulation: pesticide enclosed in capsules (or beads) of thin polyvinyl or other material, to control the rate of release of the chemical and extend the period of its activity.

environment: (1) all the organic and inorganic features that surround and affect a particular organism or group of organisms.

(2) all factors which are external to a system but which nevertheless influence that system.

EPA: Environmental Protection Agency of the United States of America. It is responsible for formation and enforcement of rules and regulations concerning pesticides in the USA

face shield: transparent piece of protective equipment used to protect the face from exposure to pesticides.

field scout: a person who samples fields to detect and record pest infestation.

first aid: emergency treatment given to an injured person before being treated by a medical doctor.

flowables: sometimes referred to as colloidal formulations, these are suspensions in a liquid or solid material in a finely divided state. Because of smallness of particular size specific gravity of the suspending medium and the presence of suspending and dispersing agents, the formulation remains almost indefinitely in a suspended stage.

fog an aerosol with particles mostly smaller than 15 μm diameter and in such large numbers that visibility is reduced.

fogger: a means of producing a fog; usually refers to a thermal system of atomisation. (see also cold fogger)

foliar application: application of a pesticide to the leaves or foliage.

formulation: way in which a technical pesticide material is prepared for practical use. Includes various solid, liquid and gaseous preparations.

fumigant: a volatile chemical (insecticide, fungicide, nematicide, etc.) that forms toxic vapours and that is used to disinfest certain areas.

fungicide: a substance that kills fungal spores or mycelium.

... eradicant: (= curative fungicide) a fungicide used to control disease after infection has occurred .

...protective: a fungicide used to protect an organism against infection by a fungus.

... systemic: a fungicide which is absorbed through a plant surface and is translocated away from the side of application.

fungistatic:	preventing the growth of a fungus without killing it.
granule:	coarse particle of inert material impregnated or mixed with an active ingredient (insecticide, herbicide or fertilizer) for application as a dry product.
harvest interval:	period between last application of a pesticide to a crop and the harvest as permitted by Jaw.
hydrophilic:	strong affinity for water.
hydrophobic:	aversion to water.
hygroscopic:	readily absorbing and retaining moisture.
inert ingredient:	any substance in pesticide product having no pesticidal action. This does not necessarily mean that it is an inactive ingredient.
insect-growth regulator:	chemical substance that disrupts the action of insect hormones controlling moulting, maturity from pupal stage to adult, and other physical processes.
integrated control:	the complementary use of biological and chemical methods to control pests (weeds, insects and pathogens.
integrated pest management:	a management system that uses all suitable techniques and methods in a manner as compatible as possible to maintain pest populations at levels below those causing economic injury. Often needed on a large area, such as an agroecosystem.
invert emulsion:	one in which the water is dispersed in oil rather than oil in water. Usually a thick viscous mixture results.
knapsack sprayer:	sprayer where the container is carried on the operator's back and held by at least two straps
lance:	a tube to hold nozzle(s) away from operator and fitted with a trigger valve.
larvicide:	chemical substance used for controlling larvae.

see median lethal concentration

see median lethal dose.

mass median diameter MMD): the figure dividing a total volume of into two equal parts; one half of the mass of the spray is contained in droplets of smaller diameter than the MMD, and the other half is contained in droplets of larger diameter.

median effective concentration (ECso): the concentration (ppm or ppb) of the toxicant that produces the designated effect in 50 per cent of the test organism exposed.

median effective dose (ED_{50}): the dose of material required to bring prescribed effect, e.g. retardation of growth in 50 per cent of the individuals in a population of a given species.

median lethal concentration (LDso): the concentration of a toxicant required to kill 50 per cent of a large sample of individuals of one species.

median lethal dose (LD_{50}): the dose of a toxicant killing 50 per cent of a large sample of individuals of one species.

microbial insecticide: a microorganism applied in the same way as conventional insecticides to control an existing pest population.

micron: micrometre - one millionth of a metre (1 μm = 10^{-6} m).

minimum tillage: practices which utilize minimum cultivation for seedbed preparation and may reduce labour and damage to soil structure (erosion).

molluscicide: a chemical used to kill or control snails and slugs.

monoculture: the continuous cultivation of either one crop species or one cultivar in both time and space.

natural control: the collective action of environmental factors, physical and biotic, to maintain over a period of time the numbers of a population within certain upper and lower limits.

natural enemies: an animal or plant which kills or injures another animal or plant, e.g. predators and parasites (insects or diseases) that attack pests.

negligible residue: a tolerance which is set on a food or feed crop permitting an ultra-small amount of pesticide at harvest as a result of indirect contact with the chemical.

nematicide: chemical used to kill nematodes.

nematode: generally microscopic, worm-like animals that live saprophytically in water or soil or as parasites of plants and animals.

nozzles: devices which atomize a spray liquid

nuclear polyhydrosis . virus (NPV): a disease-causing virus of insects; can be cultured commercially and sold as a biological insecticide.

number median diameter: half the number of droplets in a sample of spray are larger than the number median cliameter and the half of the droplets by number are smaller.

nymph: stage where the immature insects resemble the adults but are smaller and lack full development of the wings. eg grasshoppers.

operating pressure: pressure at which the spray liquid should enter the nozzle.

oral toxicity: toxicity of a compound when given by mouth;

orifice: the opening or hole in a nozzle through which a spray liquid is forced out and broken up into droplets. The size of the orifice controls the volume, whereas the shape of the orifice controls the spray pattern.

pelleting: coating of seed with inert material, often incorporating pesticides, to ensure uniform size and shape.

penetrant: an additive or adjuvant which enables the pesticide to move through the outer surface (cuticle) of plant tissues.

permitted tolerance:	the maximum content of toxicant allowed in foodstuffs for human consumption.
pest:	(1) an animal (commonly insect, mite, nematode) which causes damage to, or destruction of, plant parts, plant products, etc. (2) a plant parasite eating the host by means of mouth parts. (3) a weed
pest resurgence:	the rapid reappearance of a pest population in injurious numbers, usually brought about after the application of a broad-spectrum pesticide has killed the natural enemies which normally keep the population of a pest in check.
pesticide:	an "economic poison" used for preventing, destroying, repelling, or mitigating any pest, pathogen or weed.
pheromones:	highly potent chemical produced by organisms, which affect their behaviour; eg insect sex attractants produced by insects.
phytosynthesis:	the process by which carbon dioxide and water are combined in the presence of light and chlorophyll to form carbohydrate.
phytotoxic:	toxic to plants or plant growth.
population dynamics:	the term covers the normal changes which occur in both populations of organisms eg those associated with growth, reproduction and dissemination.
post emergence:	applied after the emergence of a specified weed or crop.
ppb:	parts per billion (parts in 10^9 parts).
ppm:	paris per million (parts in 10^6 parts).
pre emergence:	applied prior to emergence of specified weed or crop.
precipitate:	a solid substance that forms in a liquid and settles out.

pre-plant application: treatment applied on the soil surface before seeding or transplanting.

pressure pack: a small container with a pressurized liquid gas that is used to dispense aerosols.

propellant: liquified gas in a pressurized pesticide container that disperses the pesticide formulation.

protectant: substance applied to plant surface to prevent penetration and subsequent infection by an organism.

pupa: (plural: pupae) the stage between larva and adult in insects undergoing complete metamorphosis, a non-feeding and usually inactive stage.

quarantine: all operations associated with prevention of dissemination of unwanted organisms within and between territories.

random sampling: the most commonly used method of scouting for insect numbers and damage whereby samples are taken without plan (haphazardly).

ratoon: a crop which is produced by allowing a harvested crop to grow again.

re-entry (intervals): waiting interval required by law between application and entrance of workers into the treated crop without protective clothing.

registration: acknowledgement of approval of a pesticide by a legislative body concerning the uses stated on the label.

repellent: substance used to repel insects, rodents, etc.

residual herbicide: a herbicide that persists in the soil and injures of kills germinating weed seedlings over a relatively short period of time.

residue: trace of a pesticide and its metabolites remaining on and in crop tissue, soil, water or other parts of the environment after a given time.

... maximum ...limit: the maximum concentration (in mg/.kg or ppm) of a residue that is legally permitted in or on food at harvest or at a specified post-harvest processing stage up to the point of consumption.

resistance: (1) the ability of an organism to withstand or oppose the operation or to lessen or overcome the effects of an injurious or pathogenic factor.

(2) the ability of a host to suppress or retard the activity of a pathogenic organism or virus.

(3) natural or genetic ability of an organism to tolerate the poisonous effect of a toxicant.

... acquired: non-inherited resistance conferred by some predisposing treatment.

... horizontal: resistance which is more evenly spread against all races of the pathogen (equivalent to race-non-specific resistance) .

... race ... specific: resistance to some races of the pathogen,

... vertical: resistance to some races of a pathogen but not to others (equivalent to race-specific resistance).

restricted use pesticide: a pesticide that should be applied only by ce1 1ified applicators, because of their inherent toxicity or potential hazard to the environment.

retention period: the time following acquisition feeding during which a vector is capable of transmitting a virus.

RPM: revolutions per minute.

roguing: the critical examination of a crop and removal of unhealthy or otherwise unwanted plants, e.g. those that are not true to type.

run-off: (1) the process of shedding spray from plant surfaces during and immediately after application.

(2) the quantity of spray which runs off a unit area of plant surface.

safener: a chemical that reduces the toxicity of another chemical.

secondary pest: a pest which usually does little if any damage, but which can become a serious pest under certain conditions, e.g. when insecticide application destroys the pest's predators and parasites.

seed dressing: the process of cleaning seed; also used to refer to the coating of seed with pesticides.

seed treatment: the process of coating or impregnating seeds with a fungicide, etc.

senescence: process or state of growing old.

sequential sampling: a sampling system that allows a decision to be made without necessarily completing the whole sample.

signal word: a required word that appears on every pesticide label to denote the relative toxicity of the product. The signal words are *Danger-Poison* for highly toxic compounds, *Waming* for moderately toxic compounds, or *Caution* for slightly toxic ones.

slow-release formulation: formulation from which the active ingredient dissipates slowly and thus remains effective over a long period of time.

slurry: a semi-fluid mixture of solid particles in a liquid, usually water; fungicides are applied to seeds as slurries to produce a thick protective coating.

soil incorporation: mechanical mixing of herbicide with thesoil.

split treatment: when the total dosage is divided and applied on two or more occas10ns

spray: (I) to apply (water, pesticides, etc.) in the form of droplets suspended in the air.

(2) a solution or suspension (usually in water) dispersed under air- or water -pressure as droplets.

... angle: the angle measured in degrees between the two edges of a spray sheet.

... quality: the range of droplets within a spray of certain sizes, thus fine spray has a higher proportion of small droplets than a coarse spray.

sprayer: a machine for producing and applying a spray.

... air-carrier: a machine employing air to propel the spray.

... hydraulic: a machine depending on liquid energy both to form and to propel the spray.

..logarithmic: equipment fo applying a spray in concentrations decreasing exponentially with time or distance.

spreader: ingredient added to a spray mixture to assist in its even distribution over the target or plant surface.

sticker: ingredient added to spray or dust to improve its adherence to plant surfaces.

stomach poison: a pesticide that must be eaten by an insect or other animal in order to kill or control it.

stratified sampling: a system of sampling where a field is divided into sections and each part is sampled independently.

suction hose: hose used to pull water from a stream or pond or other sources into the sprayer tank, or to draw liquid from the tank into the sprayer pump.

surface tension: the phenomenon by which a drop of liquid tends to contract rather than to spread out as a film.

surfactant: abbreviated form of "surface active agent"; ingredient that enhances the surface-modifying properties of a pesticide formulation, e.g. by affecting surface tension, (wetting agent, emulsifier, etc.).

susceptible: non-immune, i.e. subject to infection.

susceptibility: inability of an organism (host) to oppose the operation or to overcome the effects of an injurious or pathogenic factor.

suspension: finely divided solid particles dispersed in a liquid.

swirl plate: disc with holes cut at an angle to cause liquid to flow in a circular direction.

symbiosis: a mutually beneficial association of two different kinds of orgarnsm.

symptom: a visible or otherwise detectable abnormality arising from a disease.

systemic: (I) chemical compound that is absorbed and translocated throughout the plant or animal, e.g. systemic fungicide.

(2) (of a plant pathogen) occurring throughout the plant.

tailboom: vertical boom with one, two or more nozzles mounted to the rear of the operator to avoid walking into spray.

tank mix: mixture of two or more pesticides in a spray tank at the time of application.

teratogenic: (I) substance that causes physical birth defects in the offspring.

(2) causing gross structural abnormalities, e.g. witches' brooms.

tolerance: (I) the ability of an affected organism to endure invasion by a pathogenic organism or virus with little or no reaction, as shown by the more or less complete absence of symptom expression and damage.

(2) amount of a pesticide residue permitted to remain on or in a crop at harvest or in processed produce. Expressed as parts per million (ppm).

toxicant: a poisonous substance such as the active ingredient in pesticide formulations that can injure of kill organisms.

toxicity: the ability to act as a toxin or toxicant.

toxin: a naturally occurring poison produced by plants, animals and microorganisms, e.g. snake venom, botulism toxin and mycotoxins.

trade name (trademark): name given by its manufacturer or formulator, distinguishing it as being produced or sold exclusively by that company.

translocation: transfer of nutrients or other substances such as a herbicide from one plant part to another.

transpiration: the loss of water vapour from the surface of leaves.

trigger valve: hand-operated valve usually mounted on a lance or in hose line between the pump and boom; usually contains a filter.

trivial name: name in general usage; e.g. nicotine, c.f. common name.

ultra-Iow volume (ulv): < 5 litres/hectare; also a type of pesticide formulation as supplied by the manufacturer.

ultraviolet (UV): that portion of the light spectrum beyond the violet band that is invisible to man , but both visible and highly attractive to nocturnal insects. Also used to trace spray droplets, containing a suitable fluorescent tracer.

US gallon 3.785 litres (Note Imperial gallon is 4.55 litres

vapour (vapor): steam, gaseous form of a normally liquid or solid substance.

very-low volume: usually 5- 20 litres of spray per hectare.

viscosity: the property of stickiness of fluids and gases that hinders flowing; power of resisting a change in the arrangement of the molecules.

volatilize: to vapourize.

volume median diameter: a measure of the droplet spectrum in which half the volume contains droplets which are larger than the VMD, and half the spray volume contains droplets which are smaller.

wettable powder: pesticide formulation of toxicant mixed with inert dust and a wetting agent that mixes readily with water and forms a short-term suspension (requires tank agitation).

wetting agent: compound that causes spray solutions to contact plant surfaces more thoroughly by decreasing the contact angle of the liquid.

withholding period: the recommended minimum period that should elapse between the last application of a pesticide to any crop, pasture or animal and the harvesting, grazing, cutting or slaughtering thereof, or the collection of milk and eggs for human consumption.

BIBLIOGRAPHY

Anon (1986) *International Code of Conduct on the Distribution and Use of Pesticides.* F AO, Rome.

Anon (1987) *Guidelines for the avoidance, limitation and disposal ofpesticide waste on thefarm.* GIF AP, Brussels.

Anon (1989) *Guidelines for personal protection when using pesticides in hot climates.* GIF AP, Brussels.

Anon (1990) *Equipment for Vector Control.* World Health Organisation, Geneva.

Bache, D. and Johnstone, D.R. (1992) *Microclimate and spray dispersion.* Ellis Horwood, Chichester.

Matthews, G.A. (1992) *Pesticide Application Method,;;.* Second Edition Longman, Harlow.

Matthews, G.A and Hislop, E.C. (eds.) (1993) *Application Technology for Crop Protection.* CABI, Wallingford.

Oudejans, J.H. (1991) *Agro-Pesticides: Properties and Fu l lctio l lsin !l ltegratedCrop Protection* ESCAP, Bangkok.

Thornhill, E.W. (1984) Maintenance and Repair of Spraying Equipment. *Tropical Pest Manageme l lt*30, 266-281.

Thornhill, E.W. (1985) A Guide to Knapsack Selection. *Tropical Pest lvfa l lageme l lt* 31, 11-17.

Acknowledgements

The material in the appendices based on BS 4115:1993 and BS 7411:1991 is reproduced with the permission of the British Standards Institution. Full copies of both standards can be obtained from BSI sales, Linford Wood, Milton Keynes, MK 14 6LE, UK; or through national standards bodies.

The reproduction of standards published by WHO is with the permission of the World Health Organisation.

Appendices

Appendix 1 Specification for Compression Sprayers

The following is based on the standard, developed in conjunction with the International Pesticide Application Research Centre and published by the British Standards Institute as BS 4115: 1993. A more detailed specification for compression sprayers was published by the World Health Organisation as WHO/VBC/89.970.

1. Scope

This Standard specifies requirements for the design, materials and performance of compression knapsack sprayers used to apply approved pesticides. It applies to sprayers of over 5 l capacity.

NOTE Sprayers of 5 l or less, i.e. garden sprayers, are outside the scope of this standard.

2. Definition

For the purposes of this British Standard, the following definition applies.

2.1 Compression knapsack sprayer

A device which consists of a tank, equipped with a hand-operated air pump for pressurization. Air pressure in the tank forces the liquid through a hose, cut-off valve and lance to the spray nozzle.

NOTE The device is normally intended to be worn on the operator's back but may, optionally, be mounted on a trolley for transport purposes.

3. Design

3.1 The sprayer shall be designed such that it does not leak from any of its component parts under all working conditions.

When tested in accordance with A.1 and A.2, leakage from the sprayer shall not exceed the limits specified in A.1. When tested in accordance with A.3, there shall be no functional failure of any individual component. When tested in accordance with

A.4, the surface retention of any spilt liquid shall not exceed the limits specified in A.4.

3.2 The tank shall have a clearly visible, permanent indicator or a permanently marked embossed line to indicate the maximum operating capacity of liquid; this shall be not more than three-quarters of the total capacity of the tank.

3.3 The tank with all fittings shall be capable of withstanding a pressure of five times the maximum working pressure, which shall be marked clearly and permanently on the tank.

3.4 The total mass of the sprayer with fittings when filled to operating capacity shall not exceed 20 kg.

3.5 A removable strainer shall be fitted upstream of the trigger valve to ensure reliable operation of the valve.

3.6 A funnel shall be provided to fit into the tank opening. This shall have sufficient open surface area to allow liquid flow without splashing or spillage during the filling of the tank and for displaced air to escape. The strainer in the tank filler hole shall be capable of handling the entire tank contents without spillage when tested in accordance with A.S.

3.7 There shall be a device to indicate the pressure inside the tank. There shall also be a device to prevent the pressure in the tank from exceeding 7 bar (I bar = 20^5 Pa - I 0^5 N/m^2). There shall also be a device to provide constant pressure not exceeding 4 bar at the nozzle.

3.8 The tank orifice shall be designed to allow any residual pressure to escape before the lid can be removed. The orifice shall have a minimum internal diameter of 50 mm.

3.9 Threaded fittings of the pressurized part of the tank that have a diameter greater than 13 mm shall be channelled to provide a release of pressure before complete removal of the fitting in order to safeguard the operator.

3.10 The sprayer shall be so designed as to leave a maximum residue to 50 ml in the tank when exhausted after use in the normal operating position.

3.11 The pump shall be designed to achieve the maximum working pressure in the tank in not more than 50 complete strokes when filled to maximum operating capacity. The pump handle shall be so designed that, when locked in its lowest position, it provides a carrying handle.

As an option, the sprayer may be fitted with a carrying strap, which shall be made of non-absorbent material and shall be at least 50 mm wide.

When tested in accordance with A.6 the strap shall not break.

3.12 The nozzle shall consist of a common body and cap conforming to BS 6356: Part 2: 1985 or Part 6: 1992, to which a range of different-sized nozzle tips can be fitted.

NOTE A variable cone that requires adjustment by hand is not suitable.

3.13 The sprayer, whether full or empty, shall remain uprights and stable when stood on a smooth surface inclined at 5° horizontal with the lance correctly stored and with the sprayer in any orientation.

3.14 There shall be provision on the sprayer to store the lance and nozzle assembly when not spraying.

3.15 The sprayer shall have no sharp edges or projections that may mJure the operator.

3.16 Hose connections shall be re-usable without the use of sealants.

3.17 The tank outlet shall be located to withstand impact.

4. **Materials**

4.1 All parts of the sprayer shall be made of materials which shall show no signs of deterioration after being tested in accordance with A.7.

4.2 All parts of the sprayer shall be protected against rust and corrosion.

NOTE Only spray mixes recommended by the manufacturer should be used.

4.3 No wood or absorbent materials shall be used in any part of the sprayer.

4.4 Solder containing lead or tin as a major component shall not be used in the sprayer construction except on joints between the lance, nozzle body, or cut-off valve.

NOTE Strainers may be fabricated using solder providing they are replaceable items.

5. Performance

5.1 When the sprayer is tested in accordance with A.8, no fatigue failure of the tank or fittings shall occur.

5.2 When tested in accordance with A.9, the cut-off valve shall show no leakage nor loss in performance.

5.3 When tested in accordance with A.10, the whole sprayer with fittings shall show no signs of impact damage.

5.4 When tested in accordance with A. 11, the hose shall show no permanent distortion, leakage or deterioration of the hose material.

6. Marking

The tank of the sprayer shall be marked clearly and permanently with the following information:

a) the number and date of this British Standard, i.e. BS 4115: 1993 (Marking BS 4115: 1993 on or in relation to a product represents a manufacturer's declaration of conformity, i.e. a claim by or on behalf of the manufacturer that the product meets the requirements of the standard. The accuracy of the claim is solely the claimant's responsibility. Such a declaration is not to be confused with third party certification of conformity, which may also be desirable.);

b) the manufacturer's name or trade mark;

c) year of manufacture;

d) manufacturer's type number;

e) manufacturer's serial number;

f) the maximum working pressure.

Tests

A.1 Leakage test

A.1.1 *Principle*

The tank is filled with water containing a tracer dye for a stated period of time and the whole sprayer is then examined for leaks.

A.1.2 *Procedure*

Fill the tank with water and a suitable tracer dye to the recommended fill mark capacity. With the trigger valve in the closed position, place the tank on a bench 600 mm from the floor and hang the hose and trigger/lance assembly from the sprayer. Leave for 15 min and then visually examine the whole sprayer for leakage, especially noting hose connection, nozzle trigger valve and the filler hole closure. Examine all entries to the tank for hoses, pump, filler hole lid and the relief valve if fitted.

NOTE .This test should be conducted without any pressure in the tank. It should be conducted with the tank cylinder in the horizontal position. This is because some compression sprayers may leak with no pressure from the filling hole; this is especially the case with elliptical filler holes which are sealed by a cam action and rely on tank pressure for sealing.

A.1.3 *Test results*

Leakage of spray liquid in a test period shall be within the following limits:

a) a maximum of 5 ml from hose connectors, control valves on the lance and similar sites;

b) a maximum of 7 ml from the tank, lid and pressure relief valve.

A.2 Inversion leak test

A.2.1 *Procedure*

Fill the tank in accordance with A.1.2.

Invert the sprayer for 5 min and examine for any leakage.

NOTE This test should be conducted without any pressure in the tank.

A.2.2 *Test results*

Leakage of spray liquid in a test period shall be within the limits specified in A.1.3.

A.3 Hydrostatic test

A.3.1 *Principle*

The tank is filled with water containing a tracer dye and pressurized for a stated period of time. The whole sprayer is then examined for leaks.

A.3.2 *Apparatus*

A.3.2.1 *External piston pump.*

A.3.2.2 *Blanking device,* to substitute for nozzle.

A.3.2.3 *Metal safety cage.*

A.3.3 *Procedure*

Fill the tank completely with water and a suitable tracer dye. By the use of an external piston pump (A.3 .2.1) connected to the spray lance after removal of the nozzle, pressurize the test circuit to maximum working pressure. The trigger valve should be open and the nozzle replaced by a blanking device (A.3.2.2). Examine all areas for leaks after 15 min of pressurization.

After depressurizing the sprayer, remove the hand lance assembly and connect the external test pump to the sprayer hose. Remove or blank off any safety relief valves. Apply twice the maximum working pressure, plus 10% but not less than 1000 kPa, for 1 min. The tank shall be enclosed in a safety cage (A.3.2.3).

NOTE It is emphasized that this test is hydrostatic and non-pneumatic. Care is particularly important when conducting this test. If the tank is of non-metallic construction, conduct the test at a temperature of 21 °C.

A.4 Tank surface retention test

A.4.1 *Principle*

All external surfaces of the sprayer are wetted with a tracer dye for a specified period of time. the surfaces are then swabbed and the amount of dye recovered is then measured.

A.4.3 *Procedure*

With the sprayer standing on a firm horizontal surface, pour a tracer dye solution over the sprayer so as to wet all external surfaces. Wait for a period of 2 min with the sprayer standing on a clean and absorbent surface. Then swab the outer surfaces of the sprayer with an absorbent material. Determine by fluorimetry or spectrophotometry the quantity of dye absorbed by the swabs. At the same time, inoculate clean swabs with an accurately measured quantity of the tracer which is then recovered to verify the stability and recovery of the tracer from the swabs.

A.4.3 *Test results*

A sprayer shall pass the test if no more than 7 ml of tracer dye solution are recovered from the surface of the sprayer.

A.5 Strainer performance test

A.5.1 *Principle*

Liquid is poured into the tank to ascertain that the strainer in the filler hole does not cause spillage.

A.5.2 *Apparatus*

A.5.2.1 *Unprofiled bucket,* of approximately 7 I capacity.

A.5.3 *Procedure*

Pour the entire contents of the tank from an unprofiled bucket (A.5.2.1) at a rate of 0.22 l/s into the tank, through the funnel in the tank filler hole. Examine for spillage.

A.6 Strap drop test

A.6.1 *Principle*

The sprayer with filled tank is suspended by its strap and then dropped repeatedly through a specified distance to ascertain that it does not break.

A.6.2 *Apparatus*

A.6.2.1 *Horizontal har;* of 75 mm outside diameter.

A.6.3 *Procedure*

Fill the tank completely to its design capacity with water and then hang the sprayer by its strap from a solid support comprising a horizontal bar of 75 mm outside diameter (A.6.2.1), simulating its carriage on the shoulder of the operator. Then lift it a distance of 300 mm and allow it to drop and hang by the strap 25 times. Examine the strap.

A.7 Solvent compatibility test

A.7.1 *Principle*

The tank is filled with a test solution which is brought into contact with all parts of the sprayer for a specified period of time. The parts are then examined against their original specification.

A.7.2 *Materials*

A.7.2.1 *Test solution,* consisting of 40% (V/V) kerosene, 20% (V/V) toluene and 40% (V/V) xylene.

A.7.4 *Procedure*

A.7.3.1 Fill the tank with test solution (A.7.2.1). Ensure that the test solution is in contact with all parts of the sprayer as in normal use for 48 h. Replace the nozzle with a bleed valve and open the trigger valve so that air is removed and solvent contact achieved.

After cleaning the sprayer, allow to drain and dry for 24 h.

If the sprayer is supplied with a regulated delivery system, test similarly all seals, gaskets and washers which are part of that system.

A.7.4 *Test results*

All seals, gaskets and hoses shall be re-usable.

A.8 Tank fatigue and durability test

A.8.1 *Principle*

The tank with all fittings is subjected to 5000 pressure cycles at O kPa to 500 kPa or the maximum working pressure to determine that there is no failure of the tank or fittings.

NOTE Although the distortion may not be readily perceptible to the eye, a compression sprayer tank bends outwards when the tank is pressurized, returning to a normal position when pressure in the containers returns to that of the atmosphere. Repeated pressurizations results in stress reversals. Fatigue failure may result; the first evidence of this is usually seen in a broken or leaking weld or a radial crack from an opening such as that designed to receive the cover or pump.

A.8.2 *Apparatus*

A.8.2.1 *Apparatus to simulate fatigue stress,* as shown in Anon (1990).

A.8.3 *Procedure*

Connect the hose connection of the tank to the manifold on the test apparatus (A.8.2.1) by the sprayer hose.

NOTE 1 A valve is provided at each point of connection with the manifold in order that any connection may be closed off

Fill the tank completely with water and fill the manifold with water to a point within approximately 200 mm of the filler hole.

Introduce compressed air into the manifold through a pressure-regulating valve and a 3-way valve. Set a timer to open and close the 3-way valve four times/min.

NOTE 2 Compressed air enters the manifold at the selected pressure. When the. valve is de-energized, the air exhausts to atmosphere. The result is that the pressure cycles applied to the water in the manifold transmit the pressure cycle to all parts of the sprayer. Examine the tank and fittings for fatigue failure.

A.9 Cut-off valve reliability test

A.9.1 *Principle*

The cut-off valve is operated at specified rates, pressures and cycles and then examined for leaks.

A.9.2 *Apparatus*

A.9.2.1 *Test rig,* capable of regulating the operation of the cut-off valve.

A.9.3 *Procedure*

Fix the cut-off valve to a test rig (A.9.2.1) which ensures that the valve is operated no more than 15 times/min. Set the pressure of spray liquid at the nozzle to 3 bar and operate the valve for 20 000 cycles. Then inspect the valve for leaks and subject the valve to 500 operations at 4 bar. Conduct this test using water and a standard spray nozzle which has an approximately discharge rate of0.750 l/min at 3.5 bar.

Test the cut-off valve for leakage at the maximum working pressure.

A.10 Tank impact (drop) test

A.10.1 *Principle*

The sprayer with filled tank is dropped from a specified height on to a wooden platform covering the floor and then examined for damage.

A.10.2 *Apparatus*

A.10.2.1 *Wooden platform,* 50 mm thick by 800 mm square.

A.10.2.2 *Metal safety cage.*

A.10.2. *Procedure*

Place a wooden platform measuring 50 mm thick by 800 mm square (A.10.2.1) on a level floor. Surround by a metal cage (A. 10.2.2) to protect the examiner from possible injury. Fill the tank to its design capacity and pressurize to the maximum working pressure. Allow it to drop from a height of 600 mm onto the platform.

For six drops, hold the tank, base down, vertically before it is dropped. For seven drops, hold the tank horizontally on its back (as worn by the operator). Hold the base of the tank at an angle of 10° to the horizontal for six drops on one side and six drops on the opposite side, making a total of 25 drops in all.

Examine the sprayer and fittings for damage.

A.11 Hose hydrostatic test

A.11.1 *Principle*

The inside and outer surfaces of the hose are immersed in a test solution for a specified period of time and at a specified pressure and are then examined for deterioration permanent distortion or leakage.

A.11.2 *Materials*

A.11.2.1 *Test solution,* in accordance with A 7.2.1.

A.11.3 *Procedure*

Select a sample of the spray hose of minimum length 500 mm. Fill the tank with a test solution of 40% (V/V) kerosene, 20% (V/V) toluene and 40% (V/V) xylene.

Test the sample hydrostatically after immersion in the test solution (A.11.2.1) with air removed (see A.7.3), testing to the maximum test pressures of the tank for 5 mm without permanent distortion of the hose or leakage.

NOTE If the sprayer is provided with an air inlet valve of the automobile type for the purposes of supplying air. pressure to the tank from an outside high pressure source, a pressure relief valve or safety valve is required. this safety valve should be non-adjustable in order to prevent pressurization of the tank beyond the limits prescribed for safety or operational purposes.

Appendix 2 Specification for Lever-operated Knapsack Sprayers

The following specification for lever-operated knapsack sprayers is based on the British Standards Specification BS7411 developed in conjunction with the International Pesticide Application Research Centre. It is reproduced here with amendments as an example of the requirements needed to improve safety and reliability.

SPECIFICATION

1. Scope

The Standard specifies requirements for the materials, finish and performance of lever operated knapsack sprayers to apply approved pesticides.

2. Definition

For the purposes of this Standard the following definition applies.

Lever operated knapsack sprayer

An assembly consisting of a reservoir or tank for carrying spray liquid, which is supported on the operator's back by straps and which the operator pumps while spraying.

A left or right-hand lever is used to operate a diaphragm or piston pump, mounted usually inside (but in some cases, outside) the sprayer reservoir or tank. Air is retained in a small compression cylinder to force the liquid to the spray nozzle and to reduce pumping pulsations.

The tank is not under pressure but the compression cylinder is, so creating a residual pressure. The equipment is designed to enable a range of nozzles to be fitted on a hand lance.

3. General requirements

3.1 The sprayer shall not leak from any of its component parts under all working conditions. When tested in accordance with A.1 the leakage of spray liquid in a test period shall be within the following limits:-

a) a maximum of 5 ml from hose connections, control valves on the lance and similar sites;
b) a maximum of 6 ml from tank, lid and ventilation valve.

When tested in accordance with A.2 and A.3 the leakage shall not exceed 5 ml.

3.2 No part of the outer surfaces of the sprayer shall entrap spillages ofliquid. The sprayer shall have a maximum surface retention of 10 ml of liquid when tested in accordance with A 5.

3.3 Spraying shall be controlled at the hand lance. A means of controlling the spray pressure shall be provided having an accuracy of± 10%. If a pressure indicator is fitted it shall be upstream of the trigger valve.

NOTE: The pressure gauge should be to a recognized standard ___ ___ _(e.g. BS 1780).

3.4 The maximum operating pressure at the nozzle shall be limited to 4 bar. (1 bar= 10^5 Pa = 10^5 N/m^2) Any pressure adjustment device shall be external to the tank. A safety device to prevent the maximum pressure being exceeded in any part of the equipment shall be fitted and designed to divert spray liqu_id to inside the tank.

3.5 The filler hole shall be not less than 100 mm across the smallest dimension. It shall have no sharp edges and shall be provided with a lid that can be securely fitted by the gloved hand of an operator.

3.6 Any air vent in the lid or tank shall limit the escape of spray liquid to 5 ml either when tested in accordance with A.2, or if the sprayer is completely inverted.

3.7 A strainer, that is removable by gloved hands, shall be provided to fit into the tank opening. This shall have sufficient open surface area to allow liquid flow without splashing to spillage during the filling of the tank and for displaced air to escape. The strainer shall be capable of handling the entire sprayer tank contents without spillage when tested in accordance with A.4.

3.8 The tank shall have a minimum capacity of 5 l.

3.9 A device shall be fitted to indicate the volume of the contents of the sprayer tank, which shall be permanently marked at 1 l intervals to an accuracy of± 5% and shall be clearly visible.

3.10 The total weight of the sprayer when fully loaded with water shall not exceed 25 kg. The shape of the sprayer/operator interface shall be designed to avoid any discomfort.

3.11 The load bearing part of any strap shall be at least 50 mm wide. A load bearing waist strap shall also be provided. All straps and padding shall be non-absorbent.

Straps shall be fitted with quick release catches and shall be adjustable whilst the sprayer is in the working position. When tested in accordance with A.9, there shall be no visible signs of damage to the straps or their support hangers at the completion of the test.

3.12 The sprayer shall remain upright and stable when stood on a level horizontal surface or lifted by the shoulder straps during normal fitting by the operator. There shall be facilities to park the lance and nozzle assembly when not spraying.

3.13 The pump handle shall be at least 400 mm long and shall be designed for right or left hand operation. A hand-grip shall be firmly fixed to the handle and shall have a minimum diameter (or minor axis) of 25 mm and a minimumlength of 100 mm.

3.14 The minimum length of the lance from the front of the hand grip to the nozzle shall be 500 mm.

3.15 Hose connections shall be reusable.

3.16 All parts of the sprayer shall be made of materials which are unaffected by 48 h use with any approved pesticide.

3.17 Any residue left in the tank after system pump out shall not exceed 50 ml, with the sprayer set on a horizontal surface.

3.18 The manufacturer shall define an adequate cleaning routine after which no more than O.OI% of the original spray liquid shall be retained.

3.19 Any filters fitted shall be capable of being cleaned and maintained or replaced.

3.20 The compression cylinder and any hose or pipework which operates under pressure shall show no signs of leakage pressure tested to 1.5 times the maximum rated pressure.

3.21 The sprayer shall show no loss in performance over the period of the test or develop any leaks when tested in accordance with A.6 and A. I.

3.22 The cut-off valve shall show no loss in performance over the period of the test or develop any leaks when tested in accordance with A. 7 and A. I.

3.23 The sprayer tank shall show no loss in performance over the period of the test or develop any leaks when tested in accordance with A. 8 and A. I.

TESTS

Test A. Performance tests

A.1 Leakage test

A.1.1 Fill the sprayer with water and a suitable tracer dye to the recommended fill mark capacity. With the trigger valves in the closed position, place the sprayer on a bench 600 mm from the floor and hang the hose and trigger/lance assembly from the sprayer. Leave the sprayer for 15 min and visually examine for leakage (see 3.1) especially noting hose connections, nozzle trigger valve and the base of the sprayer. Examine all entries to the tank for hoses, pumps, etc.

A.1.2 Pump the sprayer to 4 bar and after 15 min do the following:-

a) Re-examine the same areas for leaks (see A.1.1).

b) Check that the trigger valve does not leak in either the open or closed position.

c) Carry out a leakage test of the lance with the trigger valve open and the nozzle blanked off.

A.2 Inversion leak test

Invert the sprayer for 5 min and check that any leakages if less than that specified in 3.1. This test is additional to the tilt test described in A.3.

NOTE: The essential component of this test is to identify potential areas of leakage that may contaminate the user or storage area.

A.3 Tilt test

Fill the sprayer with water and tilt at an angle of 90° for 5 min on each of four orthogonal axes. Check that leakage from any part of the sprayer is not more than that specified in 3.1.

A.4 Strainer performance test

Pour the entire tank contents from an unprofiled bucket at a rate of 15 l/min into the tank, through the strainer in the tank filler hole.

A.5 Surface retention test

With the sprayer standing on a firm horizontal surface, pour a tracer dye solution over the sprayer so as to wet all external surfaces. Wait for a period of 2 min with the sprayer standing on a clean and absorbent surface. Then swab the outer surfaces with an absorbent material. Determine by fluorimetry or spectrophotometry the quantity of dye absorbed by the swabs. At the same time, inoculate clean swabs with an accurately measured quantity of the tracer which is then recovered to verify the stability and recovery of the tracer from the swabs. A sprayer shall pass the test if no more than 10 ml of tracer dye solution are recovered from the surface of the sprayer.

A.6 Durability test

Securely support the sprayer by its straps on a simulated operator and operate the lever mechanically, avoiding the extreme top and bottom of each stroke. Pump water from the sprayer through a manifold with nozzles giving a minimum output of 72 l/h at 3 bar at a maximum of 35 strokes/min. The water sprayed into a separate tank shall siphon back into the sprayer under test just before the pump completely empties the sprayer tank. Continue the test for a minimum of 250 h.

A.7 Cut-off valve reliability test

Fix the cut-off valve to a special test rig which ensures that the valve is operated no more than 15 times/min. Set the pressure of spray liquid at the nozzle to 3 bar and operate the valve for 50,000 cycles. After this, inspect the valve for leaks and subject the valve to 500 operations at 4 bar. Conduct this test using a standard spray nozzle which has an approximate discharge rate of 0.750 l/min at 3.5 bar. Conduct the test using water.

Test the cut-off valve for leakage at the maximum working pressure of 4 bar.

A.8 Tank impact (drop test)

Place a wooden platform, 50 mm thick by 800 mm square, on a level floor and surrounded by a metal cage to protect the examiner from possible injury. Fill the sprayer tank completely to its design capacity with water, and allow it to drop from a height of 600 mm onto the platform. In six tests, the tank is held vertically before it is dropped; it is held horizontally on its back (as worn) in seven tests and with the base of the tank an angle of 10° for six tests on one side and six tests on the opposite side, making a total of 25 drops in all.

A.9 Strap drop test

Fill the sprayer tank completely to its design capacity with water and then hang it by its strap from a solid support comprising a horizontal bar of 75 mm outside diameter, simulating its carriage on the shoulder of the operator. Then lift it a distance of 30 cm and allow it to drop and hang by the strap 25 times.

Appendix 3 Specification for Motorised Knapsack Mistblowers

This specification was published by the World Health Organisation in *Equipment for Vector Control.*

1. Specifications

1.1 Description

The knapsack mistblower shall have an engine-driven fan or airblower to produce a high-velocity airstream into which the pesticide is metered. The engine/fan unit may be mounted in a knapsack frame that can be carried comfortably on the operator's back. There shall be no sharp edges and projections that might injure workers during normal operations. The weight of the sprayer, ready for operation, filled with fuel and insecticide formulation, shall not exceed 22. 5 kg.

1.2 Materials of construction

All materials (such as stainless steel or polypropylene) shall be resistant to chemicals used in pesticide formulations under a wide range of climatic conditions, to meet the performance criteria specified (see section 2.1).

1.3 Erigine

1.3.1 Description

A suitable lightweight single-cylinder, air-cooled, two-stroke engine (currently all commercial knapsack mistblowers are fitted with a two-stroke engine) or sufficient displacement capacity, within the ·range of 35-70 cm^3 and a minimum power output of 2 horsepower, at recommended operating speed to drive the fan unit shall be fitted with a recoil starter and a guarded exhaust, directed to the rear away from the operator and positioned on the side of the engine opposite to where manual controls are positioned. Engine capacity and power shall be specified by the purchasing agency.·

The engine shall be mounted on a carrying frame by a minimum of 3 rubber mountings. The engine in this context shall include the fan and fan casing.

A pulley wheel, protected by a removable cover, shall be provided so that the engine can be started with a rope, if the recoil starter fails.

A fuel tank with a filter (strainer) and an easily accessible outlet tap shall be positioned to avoid interference with access to other components such as the spark plug, and also to reduce the risk of fuel spillage on the engine. The tank shall have sufficient capacity to allow for one hour's continuous operation of the engine at maximum speed without refilling. A strainer shall be provided in the supply line between the fuel tank and carburettor.

A dry-type air filter shall be provided directly on the carburettor air intake. Carburettor setting screws shall be readily accessible for adjustment without requiring removal of any parts or the use of offset-type tools.

A throttle control to adjust engine speed and a stop switch shall be provided in an easily accessible and visible position. The spark plug shall be positioned to avoid flooding of the electrodes.

An engine identification plate securely fastened to the engine shall be provided to show the make, model number and type.

1.3.2 Performance

At maximum speed, the engine shall be able to pass the performance tests described in section 2.2 with respect to the following criteria:-

Fuel consumption: note more than 2.0 l/h at maximum speed as described in section 2.2.2.

Time the engine runs on a full fuel tank: not less than one hour.

Noise level: not exceeding 90 decibels (dB-A) at the operator's ear in any one measurement as described in section 2.2.3.

Durability: no operational problems during a test run of 50 hours as described in section 2.2.4.

The design of the engine shall be such that all major operations may be performed without difficulty and removal of engine from the frame shall be straightforward. The contact points and magneto shall be accessible for timing adjustment without the removal of engine, spray tank or fuel tank from the frame and without the use of special tools.

1.4 Blower

1.4.1 Description

A centrifugal or axial fan or blower shall be mounted by direct drive to the engine. The fan or blower shall be protected by a casing, Jess than 45 cm in diameter, and attached to the knapsack frame by antivibration mountings. A guard shall be positioned over the inlet part. No rivets shall be used on any part of the fan or casing.

The air output of the fan shall be ducted to the nozzle through a delivery hose, attached to the fan casing and to the nozzle tube by worm-drive clips. Part of the hose shall be sufficiently flexible to permit directional control of the nozzle without operator fatigue.

1.4.2 Performance

If an engine of 35-50 cm^3 capacity is specified, the volume of air emitted at the nozzle with the engine set at the recommended full speed shall be not less than 0.13 m^3 m/s, unless otherwise stated by the purchasing agency. If an engine of over 50 cm^3 capacity is specified, the minimal values shall be increased to 0.27 m' *Is* at 70 m/s.

1.5 Spray tank

1.5.1 General

The tank shall be pressurized to a maximum pressure of 50 kPa by air from the fan, to force liquid to the nozzle at a minimal height of 45 cm above the level in the tank. Alternatively, a pump may be used to transfer the liquid to the nozzle. The floor of the spray tank shall slope to the outlet which shall be provided with a hose nipple and filter. The filter mesh shall be determined in relation to the size of the orifice of the restrictor at the nozzle. A shut-off valve at the tank outlet can be provided an optional extra.

1.5.2 Capacity

The capacity of the insecticide tank shall not exceed 10 litres with all fittings in place and/or shall not be lower than that required to allow spraying at the lowest rate of application without refilling the engine fuel tank. The tank shall have moulded graduations at a minimum of 2-litre intervals.

1.5.3 Filler opening

The filler opening shall be not less than 95 mm in diameter if circular, or in minor axis if oval, and shall be easily sealed by a lid. The lid shall form an ai1iight seal. A basket filter with a SO-mesh screen, unless otherwise specified, shall be fitted in the tank opening.

1.5.4 Delivery hose

The hose shall be made of a pesticide-inert material. The delivery hose from the spray tank to the cut-off valve and nozzle shall not exceed 10 mm in internal diameter with a minimum wall thickness of 0.15 cm and, ideally, shall be connected to the tank without a clip. Any clip provided by the manufacturer shall be of the worm-drive type. The hose shall be readily removable to allow drainage of the spray tank.

1.6 Cut-off valve

1.6.1 Description

An easily operated positive action on/off valve shall be fitted to the spray line in a convenient location between the spray tank and nozzle. A trigger valve is preferred, in which case a locking device shall be fitted to keep the valve in the open position for long periods. A filter may be incorporated into the valve. It should be so designed that all parts are readily accessible for cleaning and replacement.

1.6.2 Performance

Liquid shall not drip or leak through the valve when subjected to 5000 cycles of operation at a pressure of 100 kPa.

1.7 Air tube

1.7.1 Description

The air tube shall be defined as the whole tube from the elbow of the fan casing to the point of attachment of the nozzle. The attachment to the fan-casing elbow shall be made of flexible material. The minimum overall length of the air tube from the fan-casing elbow to the nozzle shall be 80 cm when the tube is in the horizontal spraying position. Attachment of the flexible tube to the fan elbow and front nozzle tube shall be by clips of the worm-drive or screw type.

1.7.2 Performance

Air shall not leak through the joint when the fan is operated at maximum speed. The air tube shall be flexible and easy to handle both horizontally and vertically.

1.8 Nozzle assembly

1.8.1 Description

A gaseous or centrifugal energy nozzle shall be mounted at the distant end of the delivery hose so that droplets are entrained in the airstream and projected in line with the axis of the delivery tube. When interchangeable nozzle tips are provided, they shall be capable of being securely attached to the liquid discharge tube and be easily interchangeable. Each interchangeable nozzle tip shall be marked and/or colour-coded to indicate the diameter of its orifice.

1.8.2 Flow control

The flow rate to the nozzle shall be controlled by selecting one of a series of interchangeable restrictors to ensure a constant emission rate within± 4% of the stated amount of a reference test liquid at full engine speed. Alternatively, a variable flow regulator may be specified by the purchasing agency, but at each setting such a regulator must meet the same tolerance as interchangeable restrictors.

1.8.3 Performance

Spray droplets emitted from machines with a 35-cm^3 engine operated at specified full engine speed shall be projected not less than 10 m horizontally, as indicated by a droplet density of not less than I Oper cm_2, or 6 m veliically, when wind velocity is less than 0.5 m/s. With a 70-cm^3

Appendix 4 Specification for Hand-carried spinning disc Sprayers

At present there is no international specification for this type of equipment designed to produce a narrow droplet spectrum. The following is a guideline on the requirements for purchasing these sprayers.

1. Scope

This standard specifies requirements for the design, materials and performance of small hand-carried, spinning disc sprayers used to treat outdoor field crops with spray volumes generally Jess than 30 litres per hectare. It does not include equipment used in glasshouses.

2. Definition

A sprayer with a long handle to carry an atomizer head away from the operator's body. The atomizer head has a rotary disc that can be flat or frustoconical (cupped) in shape, and this is driven by an electric motor which is usually battery powered.

3. Design

3.1 The disc can be 1.0 - 10cm diameter; usually 5.0 - 8.0 cm. The periphery of the disc should be serrated with each point fed by a groove to dispense liquid fed to near the centre of the disc surface.

3.2 The liquid flow to the disc shall be through an interchangeable metering orifice, the size of which can be selected in relation to liquid viscosity to deliver the optimal volume for a given disc speed/ droplet size required.

3.3 The pesticide shall be contained in a bottle or tank positioned above the atomizer so that liquid is fed by gravity to the metering orifice. The maximum flow rate should be 200 ml/minute. The liquid reservoir shall have an air bleed to provide a constant head height. A separate knapsack leakproof container may be provided to refill the reservoir on the handle.

3.4 The disc shall be mounted on the shaft of a small electric motor such that the motor is protected from pesticide liquid entering it. The motor is usually a permanent magnet brushed DC type. The power consumption shall be less than 3 watts. Motor durability may be checked and shall be not less than 500 hours with the test described in 4.1.

3.5 The motor shall be connected via an on-off switch to a power supply.

3.6 The power supply may be a series of dry cells or rechargeable batteries carried either in the handle or a separate container.

3.7 The disc speed shall be suitable to produce either droplets in the 50 - 150μm VMD range for controlled downwind drift spraying usually with insecticide or fungicides, or over 250μm VMD diameter for placement of herbicides, when using an unshrouded disc. The droplet spectrum should have a VMD/NMD ratio of less than 2. Droplet spectra shall be checked using the test in 4.2.

3.8 The disc may be shrouded to restrict distribution of large herbicide droplets (>150μm VMD) to a narrow swath.

3.9 The handle shall be robust and of sufficient length to keep the atomizer at least 0.5m from the operator's body. The handle may be telescopic.

3.10 All materials of construction shall be as resistant as possible to all pesticides and solvents used in formulations for ultra-low and very-low volume applications.

3.11 The maximum weight carried in the hand shall be less than 3kg. when ready to start spraying.

3.12 The sprayer shall be supplied with an illustrated instruction manual with spare parts list.

Test Methods

4.1 Durability of motor. The atomizer head shall be mounted so that the motor can be switched on for 20 minutes and allowed to stop for 5 minutes for a total period of at least 500 hours, while water is fed to the disc at the rate of 60 ml/minute, when switched on.

4.2 Droplet sizes. The droplet spectrum shall be measured using a light diffraction laser particle size analyzer with the disc positioned to project spray droplets through the laser beam at 20 cm from the edge of the disc. The test liquid shall be water with surfactant eg. 0.1 per cent 'Agra!'.

4.3 Chemical compatability. All seals and other components that come into contact with the pesticide liquid should be immersed in a sample of the pesticide to check its compatability. The immersion should be 48 hours at 21 °C., following which the components shall be dried for 24 hours and checked for weight changes or distortions.

Appendix 5 Specification for Hand-carried Thermal Fogger

This specification was published by the World Health Organisation in *Equipment for Vector Control.*

Pulse-jet-type thermal fogger

1. Specifications

1.1 Descriptions

A thermal fog applicator shall have a system for producing a flow of hot air into which the insecticide formulation is metered. The air velocity shall be sufficient to shear the liquid into droplets and the temperature (>200°C) high enough to vaporize the droplets rapidly before the hot gas meets the surrounding air at ambient temperature.

The pulse-jet-type thermal fogger, either hand-carried or vehicle-mounted, shall have a combustion chamber in which a series of charges of fuel mixed with air are fired so that the subsequent evaluation following the firing of each charge shall escape through a pipe, the diameter of which is smaller than that of the combustion chamber, to increase the velocity of the hot gases. The insecticide formulation shall be in a container that can be pressurized to force the liquid to a nozzle (or jet) fitted into the exhaust tube. The machine shall be fitted with a safety cut-off valve to prevent the insecticide solution from flowing into the exhaust gases when the pulse-jet engine has stopped. There shall be no sharp edges or projections that might injure operators during normal operation. The weight of a hand-carried unit, ready for operation, filled with fuel and insecticide formulation, shall not exceed 16 kg. The weight of larger units, vehicle-mounted, assembled for operation, including the weight of fuel and insecticide formulation, shall not exceed 250 kg.

1.2 Materials of construction

All materials used in the construction shall be resistant, or painted or otherwise coated to provide resistance to insecticides and chemicals used in their formulations under a wide range of climatic conditions and shall meet the performance criteria specified (see section 2.1).

1.3 Pump

A simple manually operated pump or equivalent shall be provided to pressurize the insecticide and fuel containers when starting the machine. The capacity of the pump shall provide a pressure of 30-50 and 8-10 kPa respectively for the two containers. The position of the pump shall allow easy access to an electrical switch to connect a battery power supply via a spark plug fitted in the combustion chambers. (Note: this switch is only operated until the machine has started to run continuously.)

1.4 Pulse-jet engine

This shall consist of a combustion chamber with an output of not less than 9.3 kW (8000 kcal/hour). It should be fitted with a standard spark plug with an electrode gap of 2 mm, and a carburettor to meter the fuel and air flow. The fuel control valve and air intake valves shall be easily removable for cleaning. Material used in construction of the combustion chamber and exhaust tube shall be resistant to insecticide formulations and excessive heat. The exhaust tube shall be protected by a guard.

1.4.1 Performance

The engine shall pass the performance tests described m part A, section 2,2, with respect to the following criteria:

(a) fuel consumption in litres per hour as described in section 2.2.1;
(b) ability of the engine to operate for not less than one hour of fogging without the need to refuel;
(c) noise level in decibels as described in section 2.2.2;
(d) durability, with no operational problems after 50 hours as described in section 2.2.3.

1.5 Fuel tank

The capacity of the tank shall allow operation for at least one hour. A filter shall be fitted to the outlet. The fuel tank shall be protected against excessive heat from the combustion chamber and exhaust tube.

1.6 Insecticide tank

The insecticide tank shall have a capacity of not less than 5 litres. The tank opening shall be'provided with an airtight lid which is easily fitted. An outlet shall be provided to facilitate drainage. The tank shall be constructed of materials resistant to chemicals and be able to withstand without leakage a pressure at least twice the recommended operating pressure.

1.7 Metering system

The thermal fogger shall be provided with a series ofjets of different internal bore diameter to control the flow of insecticidal liquid in the range I0-30 l/hour. Larger units mounted on vehicles may have an output of up to 150 l/hour. There should also be an easily accessible on/off valve. At the request of the purchaser, an adjustable metering system may be provided an option.

1.8 Mounting frame and strap

Small hand-carried thermal foggers shall be provided with a shoulder strap. Larger machines shall be mounted on a robust frame to withstand prolonged vibration when the equipment is used on vehicles driven along rough roads and be equipped with parts or devices to reduce transmission of vibration of the vehicle to components o of the thermal fogger. The insecticide tank and fuel tank may be mounted on the same frame.

1.8.1 Straps

A chemical-resistant shoulder strap shall be provided for small hand-carried thermal foggers. The strap shall be not less than 50 mm wide over the shoulder and adjustable in length. The strap shall be fitted with shoulder-pads which shall have a minimum width of 7 cm and a minimum length of 15 cm. The strap, including the covering of any additional padding, shall be made of non-absorbent woven material. The strap shall withstand the test described in section 2.4.

1.9 Markings

The appropriate parts of the thermal fogger shall be clearly embossed with the following:

(a) capacity of insecticide tank in litres;
(b) fuel tank capacity in litres;
(c) quality of fuel to be used for the engine;
(d) open and closed position of the cut-off valve on the spray line;
(e) jet or restrict or size;
(f) setting position of the fuel valve;
(g) position of the batteries in the holder.

1.10 Spare parts and tools

The manufacturer, unless otherwise specified by the purchaser, shall supply spare parts and tools, including funnels, required for routine maintenance and replacement during normal use of the equipment over a period of one year.

A list of spare parts, illustrated and coded, shall be provided to the purchaser.

1.11 Operation and maintenance manual

An adequately illustrated manual on the pulse-jet-type thermal fogger and its maintenance and repair, in the language specified by the purchasing agency, shall be provided with each machine. The manual shall include complete instructions on the use, maintenance and cleaning of the equipment and give a list of all parts.

2. Test procedures

The test procedures recommended in these specifications are not intended to exclude accepted methods that ma be in use in different countries. However, in the event of disagreement, the procedures described herein shall govern. As far as possible, all tests should be carried out with the complete unit without dismantling any component.

2.1 Test for resistance of construction materials to chemicals

To test nonmetallic parts of the thermal fogger that are in contact with insecticide formulations, the following solvent mixture shall be used: 40% kerosene (by volume); 20% toluene (by volume); and 40% xylene (by volume). The parts being tested shall be soaked in the mixture for 72 hours at ambient temperature of at 21 °C (whichever is the highest). The parts shall then be allowed to dry in the air for a further 24 hours. After this exposure, the parts shall not deform and when they are reassembled they shall not leak and shall function normally. Care should be taken when handling this mixture as it is toxic and inflammable.

2.2 Pulse-jet engine

2.2.1 Fuel consumption

All fuel shall be drained from the tank and a measured volume of fuel shall be placed in the fuel tank. With the insecticide tank filled with kerosene or equivalent, i.e. without any active insecticide, the fogger shall be started and normal operating procedures used to determine the length of time the engine will run and the measured volume of fuel. The mean fuel consumption expressed in litres per hour shall be determined from at least 2 tests, and the time for which the machine runs on a full tank shall be calculated.

2.2.2 Noise level

The noise level shall be measured adjacent to the operator 's ear when he is in the operating position. If the noise level exceeds 90 decibels (dB-A) the purchaser shall be provided with ear protectors appropriate to the noise spectrum produced by the engine, as part of the equipment. The noise level shall never exceed 120 decibels.

2.2.3 Durability of engine

The thermal fogger shall be operated at the maximum flow rate for at least 5 hours or 5 full tanks (whichever is longer) per day. After refilling of the fuel tank, the engine shall be restarted and any difficulties in restarting and maintenance required noted. This test shall be repeated for at least 10 days or a total of 50 hours without any major interruption in the use of the machine.

2.3 Nozzle performance tests

2.3.1 Throughput

This shall be checked with the range of restrictors provided with the machine. Unless otherwise specified, odourless kerosene shall be the test liquid.

2.3.2 Droplet size

Glass slides coated with Teflon or silicon may be placed horizontally on a floor in a chamber (0.5 x 0.5 x 0.5 m) into which the fog is projected while travelling past the opening at 1 m/s and allowed to settle for at least 15 minutes. Temperature and humidity during the test shall be recorded. Other proven methods of droplet size determination may be used at the discretion of the Collaborating Centre. In particular, advanced laser systems for measuring spray droplets are available at some centres. Ideally the droplet size shall be in the range 8-20 μm VMD.

2.4 Strap test

A weight equivalent to the maximum operational weight of the hand-carried pulse-jet-type thermal fogger shall be fixed to the lower end of the strap and fitting which has been removed from the sprayer and firmly attached to a horizontal support so that the lower end is at least 20 cm above the ground. The weight shall be lifted to the height of the horizontal support and allowed to drop. The test shall be repeated 10 times. The strap shall not be broken or damaged by this test.

2.5 Field performance and durability test

The pulse-jet-type thermal fogger shall successfully undergo a field trial for a period of 250 hours without a change in the specifications provided by the manufacturers and development of major and/or frequent break-downs of any part(s) of the unit. The field trial shall be performed in accordance with the guide-lines for field trials of ground space spraying equipment.

Appendix 6 Specification for Hand~carried Aerosol Dispenser (Cold Fogger).

The hand-carried aerosol generator shall have an engine-driven air compressor to produce a flow or vortex of air into which the insecticide formulation is metered so that the liquid is sheared into droplets no larger than 25µm VMD. The unit shall be provided with a handle that is mounted so that the weight is evenly distributed. The complete unit assembled for operation shall not weigh more than 12 kg.

Materials ofconstmction

All materials used in construction shall be resistant to pesticides and other chemical used in their formulations under a wide range of climatic conditions.

Dimensions

The aerosol generator shall not exceed overall dimensions of 60 x 40 x 30 cm. The machine shall be easy to hold by a handle that does not get too hot during use.

Engine

The unit shall be equipped with either an electric AC motor of watts. (Voltage as specified by purchaser) or a 2-stroke combustion engine. The noise level shall not exceed 86 decibels, otherwise hearing protection shall be provided.

Blower

The air supply shall be provided by a fan to deliver a minimum of $0.5m^3$ per minute to the nozzle.

Nozzle

The nozzle shall be designed to shear the insecticide liquid delivered at a maximum of 50 ml/minute into aerosol droplets (VMD <25µm).

Insecticide tank

A container of at least 1 litre capacity shall be provided. Where possible insecticide shall be pre-packed in containers that fit the equipment directly without the need to transfer the chemical to the aerosol generator container provided with the equipment.

Metering system

The insecticide container shall have a metering valve incorporated in the design, or a valve shall be part of the nozzle design. The equipment shall have an easily operated on/off valve.

Markings

Operating instructions shall be clearly indicated on the machine. There shall also be a maintenance schedule given in an operator's manual, which specifies the spare parts and tools that are likely to be required.

CONVERSION TABLES

	A	B	A - B	B - A
Weight	oz	g	x28.35	x0.0353
	lb	kg	x0.454	x2.205
	cwt	kg	x50.8	x0.0197
	ton(long)	kg	x1016	x0.000984
	ton(short)	ton(long)	x0.893	x1.12
Surface area	in^2	cm^2	x6.45	x1.155
	ft^2	m^2	x0.093	x10.764
	yd^2	m^2	x0.836	x1.196
	yd^2	acres	x0.000207	x4840
	acres	hectares	x0.405	x2.471
Length	µm	mm	x0.001	x1000
	in	cm	x2.54	0.394
	ft	m	x0.305	x3.281
	yd	m	x0.914	x1.094
	mile	km	x1.609	x0.621
Velocity	ft/s	m/s	x0.305	x3.281
	ft/min	m/s	x0.00508	x197.0
	mile/h	km/h	x1.609	x0.621
	mile/h	ft/min	x88.0	x0.0113
	knot	ft/s	x1.689	x0.59
	m/s	km/h	x3.61	x0.277
	cm/s	km/h	x.0.036	x27.78
Quantities/ area	lb/acre	kg/ha	x1.12	x0.894
	lb/acre	mg/ft^2	x10.4	x0.09615
	kg/ha	mg/m^2	x100	x0.01
	mg/ft^2	mg/m^2	x10.794	x0.093
	oz/yd^2	cwt/acre	x2.7	x0.37
	gal(Imp.)/ acre	litres/ ha	x11.23	x0.089
	gal(USA)/ acre	litres/ ha	x9.346	x0.107
	fl.oz(Imp)/ acre	ml/ha	x70.05	x0.0143
	fl.oz(USA)/ acre	ml/ha	x73.14	x0.0137
	oz/acre	g/ha	x70.051	x0.0143
	oz/acre	kg/ha	x0.07	x14.27

Dilutions	fl.oz/100 gal(Imp)	ml/100 litres	x6.25	x0.16
	pint/100 gal(Imp)	ml/100 litres	x125	x0.008
	oz/gal (Imp)	g/litre	x6.24	x0.16
	oz/gal(USA)	g/litre	x7.49	x0.134
	lb/10 gal (Imp)	kg/100 litres	x0.0998	x10.02
Density of water				
	gal(Imp)	lb	x10	x0.1
	gal(USA)	lb	x8.32	x0.12
	lb	ft^3	x0.016	x62.37
	litre	kg	x1	x1
	ml	g	x1	x1
	lb/gal (Imp)	g/ml	x0.0997	x10.03
	lb/gal (USA)	g/ml	x0.1198	x8.34
	lb/ft^3	kg/m^3	x16.	x0.0624
Volume	in^3	ft^3	x0.000579	x1728
	ft^3	yd^3	x0.037	x27
	yd^3	m^3	x0.764	x1.308
	fl.oz(Imp)	ml	x28.5	x0.0352
	fl.oz(USA)	ml	x29.6	x0.0338
	gal(Imp)	gal(USA)	x1.20	x0.833
	gal(Imp)	litre	x4.55	x0.22
	gal(USA)	litre	x3.785	x0.264
	cm^3	m^3	$x10^{-6}$	$x10^{6}$
	cm^3	μm^3	$x10^{12}$	$x10^{-12}$
Pressure				
	lb/in^2	kg/cm^2	x0.0703	x14.22
	lb/in^2	bar	x0.0689	x14.504
	bar	kPa	x100	x0.01
	lb/in^2	kPa	x6.89	x0.145
	kN/m^2	kPa	x1	x1
	N/m^2	kPa	x0.001	x1000
	lb/m^2	atm	x0.068	x14.696
Power				
	hp	kW	x0.7457	x1.341
Temperature				
	°C	°F	9/5 °C+32	5/9(°F-32)

Pesticide calculations

1. To determine the quality (X) required to apply the recommended amount of active ingredient per hectare (A) with a formulation containing *B* percentage active ingredient.

$$\frac{A \times 100}{B} = X$$

Example: Apply 0.25 kg ai/ha of 5 per cent carbofuran granules

therefore $\frac{0.25 \times 100}{5} = 5$ kg granules/ha

2. To determine the quantity of active ingredient (Y) required to mix with a known quantity of diluent (Q) to obtain a given concentration of spray

$$Q \times \frac{\text{per cent concentration required}}{\text{per cent concentration of active ingredient}} = Y$$

(a) Example: Mix 100 litres of0.5 per cent ai, using a 50 per cent wettable powder

$$\frac{100 \times 0.5}{50} = 1 \text{ kg ofwettable powder}$$

(b) Example: Mix 2 litres of 5 per cent ai using a 75 per cent wettable powder

$$\frac{2000 \times 5}{75} = 133 \text{ g of wettable powder}$$

Units abbreviations and symbols

Units

A	ampere
atm	atmospheric pressure
bar	barometric pressure
cm	centimetre
dB	decibel
fl. oz	fluid ounce *
g	gramme
g	acceleration due to gravity (9.8 mJs^{-2})
gal	gallon*
h	hour
ha	hectare
hp	horsepower
kg	kilogram
km	kilometre
kN	kilonewton
kPa	kilopascal
kW	kilowatt
l	litre
m	metre
mg	milligramme
ml	millilitre
mm	millimetre
µm	micrometre
N	newton
µP	micro-poise
p	poise
pst	pounds per square inch
pt	pint
pto	power take-off
s	second
V	volt

*Volume measurements may be in Imperial or American units as indicated by (Imp.) or (USA).

A	area
a	average distance between airstrip or water supply to fields
ac	alternating current
adv	average droplet volume
ai	active ingredient
AN	Antanov aircraft
BPMC	fenobucarb
C	average distance between fields
CDA	controlled droplet application
D	diameter of centrifugal energy nozzle or opening of nozzle
d	droplet diameter
DCD	disposable container dispenser
'D' cell	a standard size dry battery
de	direct current
DUE	deposit per unit emission
ec	emulsifiable concentrate
EDX	energy dispersive X-ray
EPA	Environmental Protection Agency (USA)
F	average size of field
FAQ	Food & Agriculture Organization of the United Nations
FN	flow number
FP	fluorescent particle
GIFAP	Groupement International des Associations Nationales de Fabricants de Produits Agrochimiques (International Group of National Associations of Manufacturers of Agrochemical Products)
H	height
HAN	heavy aromatic naphtha
HCN	hydrogen cyanide
HLB	hydrophile-lipophile balance
HP	high power battery
HV	high volume
Hz	hertz
ID	internal diameter
IGR	insect growth regulator
IPM	integrated pest management
IRM	insecticide resistance management
ISA	international standard atmosphere
K,k	constant
kV	kilovolt
L	length
LAI	leaf area index
LD_{50}	median lethal dose
LV	low volume

MCPA	4-chloro-o-tolyloxyacetic
MRL	maximum residue level
MV	medium volume
N,n	number of droplets
nmd	number median diameter
NPV	nuclear polyhedrosis virus
OES	occupational exposure standares
p	particle parameter
PIC	prior informed consent
PMS	particle measuring system
PRY	pressure regulating valve
PTFE	polytetrafluoroethylene
PVC	polyvinyl chloride
Q	application rate (litre/ha)
q	application rate (litre/rn^2)
Q.	volume of air
Qf	quantity of spray per load
qn	throughput of nozzle
Qt	volume aplied per minute
rev	revolution
rpm	revolutions per minute
s	swath
s	distance droplet travels
SP	single power battery
SMV	spray management valve
SR	stability ratio
T	temperature
I;	time per loading and turning
J;v	turn time at end of row
U,u	wind speed
ULV	ultra low volume
UR	unsulfonated residue
UV	ultra-violet light
V	velocity
Vr	velocity of sprayer while ferrying
Ys	velocity of sprayer while spraying
vad	volume average diameter
VLV	very low volume
vm	volume median diameter
VRU	variable restrictor unit
w	width
w	angular velocity
WHO	World Health Organization
wp	wettable powder

surface tension
viscosity of air
density of air
density of droplet
in less than
greater than

FAQ TECHNICAL PAPERS

FAO AGRICULTURAL SERVICES BULLETINS

	Farm planning in the early stages of development, 1968 (E Rep. 1982 - F S)
2	Planning for action in agricultural development, 1969 (E F S)
3	Karakul processing, 1969 (E)
4	Bread from composite flour, 1969 (E • F S)
5	Sun-drying of fruits and vegetables, 1969 (E F S)
6	Cashew nut processing, 1969 (E F S)
7	Technology for the production of protein foods from cottonseed flour, 1971 (E F S)
8	Cassava processing, 1971 (New edition, 1977, available in E, F and S in the FAO Plant Production and Protection Series, No. 3)
9	Worldwide list of food technology institutions, 1971 (E/F/S*)
10	Technology of production of edible flours and protein products from groundnuts, 1971 (E F S)
11	Technology of production of edible flours and protein products from soybean, 1971 (E F S)
12	A guide for instructors in organizing and conducting agricultural engineering training courses, 1971 (E F S)
12 Sup. 1	Elements of agricultural machinery, Vol. 1, 1977 (ES)
12 Sup. 2	Elements of agricultural machinery, Vol. 2, 1977 (ES)
13	Fruit juice processing, 1973 (E S)
14	Environmental aspects of natural resource management - agriculture and soils, 1972 (E F S)
15	Manual on sericulture: Vol. I - Mulberry cultivation, 1976 (E F) Vol. 2 - Silkworm rearing, 1973 (E F) Vol. 3 - Silk reeling, 1972 (E F)
16	The use of aircraft in agriculture, 1972 (New edition, 1974, available in E, F and Sin the FAO Agriculture Series, No. 2)
17	Airtight grain storage, 1973 (E F S) ,
18	Rice testing methods and equipment, 1973 (C E)
19	Cold storage - design and operation, 1973 (E F S)
19/2	Design and operation of cold stores in developing countries, 1984 (Ar E F S)
20	Processing of natural rubber, 1973 (E)
21 Rev. 1	Agricultural residues: world directory of institutions, 1978 (E/F/S)
21 Rev. 2	Agricultural residues: world directory of institutions, 1982 (E/F/S)
22	Rice milling equipment operation and maintenance, 1974 (C E)
23	Rice drying (E**)
24	Worldwide list of textile research institutes, 1974 (E/F/S)
25	Molasses utilization, 1975 (E F S)
26	Tea processing, 1974 (E)
27	Some aspects of earth-moving machines as used in agriculture, 1975 (E)
28	Mechanization of irrigated crop production, 1977 (E)
29	Non-mulberry silks, 1979 (E)
30	Machinery servicing organizations, 1977 (E)
31	Rice-husk conversion to energy, 1978 (E)
32	Animal blood processing and utilization, 1982 (C ES)
33	Agricultural residues: compendium of technologies, 1978 (E/F/S)
33 Rev. 1	Agricultural residues: compendium of technologies, 1982 (E/F/S)
34	Farm management data collection and analysis, 1977 (E F S)
35	Bibliography of agricultural residues, 1978 (E/F/S)
36	China: rural processing technology, 1979 (E)
37	Illustrated glossary of rice-processing machines, 1979 (Multil)
38	Pesticide application equipment and techniques, 1979 (E)
39	Small-scale cane sugar processing and residue utilization, 1980 (E F S)
40	On-farm maize drying and storage in the humid tropics, 1980 (C E)
41	Farm management research for small farmer development, 1980 (C E F S)
42	China: sericulture, 1980 (E)
43	Food loss prevention in perishable crops, 1981 (E F S)
44	Replacement parts for agricultural machinery, 1981 (E F)
45	Agricultural mechanization in development: guidelines for strategy formulation, 1981 (E F)
46	Energy cropping versus food production, 1981 (E F S)
47	Agricultural residues: bibliography 1975-81 and quantitative survey, 1982 (E/F/S)
48	Plastic greenhouses for warm climates, 1982 (E)
49	China: grain storage structures, 1982 (E)
50	China: post-harvest grain technology, 1982 (E)
51	The private marketing entrepreneur and rural development, 1982 (E F S)
52	Aeration of grain in subtropical climates, 1982 (E)
53	Processing and storage of foodgrains by rural families, 1983 (E F S)
54	Biomass energy profiles, 1983 (E F)
55	Handling, grading and disposal of wool, 1983 (Ar E F S)
56	Rice parboiling, 1984 (E F)
57	Market information services, 1983 (E F S)
58	Marketing improvement in the developing world, 1984 (E)
59	Traditional post-harvest technology of perishable tropical staples, 1984 (E F S)
60	The retting of jute, 1985 (E F)
61	Producer-gas technology for rural applications, 1985 (E F)
62	Standardized designs for grain stores in hot dry climates, 1985 (E F)
63	Farm management glossary, 1985 (E/F/S)
64	Manual on the establishment, operation and management of cereal banks, 1985 (E F)
65	Farm management input to rural financial systems development, 1985 (E F S)
66	Construction of cribs for drying and storage of maize, 1985 (E F)
67	Hides and skins improvement in developing countries, 1985 (C E F)
68	Tropical and sub-tropical apiculture, 1986 (E)
68/2	Honeybee mites and their control - a selected annotated bibliography, 1986 (E)
68/3	Honey and beeswax control, 1990 (E** S)
68/4	Beekeeping in Asia, 1986 (E)
68/5	Honeybee diseases and enemies in Asia: a practical guide, 1987 (E)
68/6	Beekeeping in Africa, 1990 (E)
69	Construction and operation of small solid-wall bins, 1987 (E)

70 Paddy drying manual, 1987 (E)
71 Agricultural engineering in development: guidelines for establishment of village workshops, 1988 (C E F)
72/1 Agricultural engineering in development - The organization and management of replacement parts for agricultural machinery, Vol. 1, 1988 (E)
72/2 Agricultural engineering in development - The organization and management of replacement parts for agricultural machinery, Vol. 2, 1988 (E)
73/1 Mulberry cultivation, 1988 (E)
73/2 Silkworm rearing, 1988 (E)
73/3 Silkworm egg production, 1989 (E)
73/4 Silkworm diseases, 1991 (E)
74 Agricultural engineering in development: warehouse technique, 1989 (E F S)
75 Rural use of Jignocellulosic residues, 1989 (E)
76 Horticultural marketing - a resource and training manual for extension officers, 1989 (E F S)
77 Economics of animal by-products utilization, 1989 (E)
78 Crop insurance, 1989 (E S)
79 Handbook of rural technology for the processing of animal by-products, 1989 (E)
80 Sericulture training manual, 1990 (E)
81 Elaboraci6n de aceitunas de mesa, 1991 (S)
82 Agricultural engineering in development: design and construction guidelines for village stores, 1990 (E F S)
83 Agricultural engineering in development: tillage for crop production in areas of low rainfall, 1990 (E)
84 Agricultural engineering in development: selection of mechanization inputs, 1990 (E F S)
85 Agricultural engineering in development: guidelines for mechanization systems and machinery rehabilitation programmes, 1990 (E)
86 Strategies for crop insurance planning, 1991 (ES)
87 Guide pour l'etablissement, Jes operations et la gestion des banques de cereales, 1991 (F)
88/1 Agricultural engineering in development - Basic blacksmithing: a training manual, 1991 (ES)
88/2 Agricultural engineering in development - Intermediate blacksmithing: a training manual, 1991 (E F S)
88/3 Agricultural engineering in development - Advanced blacksmithing: a training manual, 1991 (E F)
89 Post-harvest and processing technologies of African staple foods: a technical compendium, 1991 (E)
90 Wholesale markets - Planning and design manual, 1991 (E)
91 Agricultural engineering in development: guidelines for rebuilding replacement parts and assemblies, 1992 (ES)
92 Agricultural engineering in development: human resource development - training and education programmes, 1992 (E F S)
93 Agricultural engineering in development: post harvest operations and management of foodgrains, 1994 (E F S)
94 Minor oil crops:
Part I - Edible oils
Part II - Non-edible oils
Part III - Essential oils, 1992 (E)
95 Biagas processes for sustainable development, 1992 (E F)
96 Small-scale processing of microbial pesticides, 1992 (E)
97 Technology of production of edible flours and protein pro.ducts from soybeans, 1992 (E F)
98 Small-, medium- and large-scale starch processing, 1992 (E F)
99/1 Agricultural engineering in development: mechanization strategy formulation - Vol. 1, Concepts and principles, 1992 (E F S)
100 Glossary of terms for agricultural insurance and rural finance, 1992 (E)
101 Date palm products, 1993 (E)
102 Experiencias de mercadeo de pequefios agricultores en el marco de proyectos de desarrollo rural, integrado, 1992 (S)
103 Banking for the environment, 1993 (E S)
104 Agricultural engineering indevelopment: agricultural tyres, 1993 (E)
105 Apicultura practica en America Latina, 1993 (S)
106 Promoting private sector involvement in agricultural marketing in Africa, 1993 (E F)
107 La comercializaci6n de alimentos en los grandes centros urbanos de America Latina, 1993 (S)
108 Plant tissue culture: an alternative for production of useful metabolites, 1993 (E)
Grain storage techniques - Evolution and trends in developing countries, 1994 (E)
110 Testing and evaluation of agricultural machinery and equipment - Principles and practices, 1994 (ES)
111 Low-cost, urban food distribution systems in Latin America, 1994 (E S)
112/1 Pesticide application equipment for use in agriculture - Vol. 1, Manually carried equipment, 1994 (E)
113 Maintenance and operation operation of bulk grain stores, 1994 (E)

Availability: October 1994

Ar - Arabic
C - Chinese
E - English
F - French
P - Portuguese
S - Spanish
Multil - Multilingual
Out of print
In preparation

The FAO Technical Papers are available through the authorized FAO Sales Agents or directly from Distribution and Sales Section, FAO, Viale de/le Terme di Caracalla, 00100 Rome, Italy.

This bulletin describes the main types of manually operated or manually carried pesticide application equipment. It gives deta.ils of design and shows the types of nozzle needed to treat crops and to improve the safety and efficiency of the spraying operation. Besides the popular lever-operated knapsack sprayers and compression sprayers, motorized knapsack sprayers, spinning disc applicators for reduced volume and special equipment such as fog, granule and dust applicators are included. Specifications for spraying equipment are provided as well as g1.1idanceon safety and maintenance. The bulletin is directed towards agricultural engineers, technicians, training establishments and others interested in the subject.

It provide$ practical guidance for those involved in pesticide application.